Beyond Posture

Beyond Posture

Why Security Must Govern Trajectories, Not Snapshots

HIMANSHU TIWARI

Building Security for an Autonomous World

Printed in the United States of America
Library of Congress control number (LCCN): 2026915247

First edition.

This book is for educational and informational purposes only and does not constitute legal, financial, compliance, security, consulting, or professional advice. The examples and frameworks are illustrative and do not guarantee any specific outcome.

The views expressed are solely those of the author and do not represent any current or former employer, client, institution, or affiliated organization.

For permissions or inquiries, contact: contact@thefifthdraft.com
ISBN: 979-8-9962147-0-9 (hardcover)
ISBN: 979-8-9962147-1-6 (paperback)]
ISBN: 979-8-9962147-2-3 (Kindle)

To my family, for their unconditional support and for inspiring me to pursue my own trajectory … my bedrock.

I remain profoundly grateful for my time at Cigital—a company way ahead of its time—and for the lifelong connections I made there. Rarely do I come across a mature software security program that hasn't been shaped by its influence or led by an alum … The Cigital Mafia!

Table of Contents

INTRODUCTION

The Moment We're In

A security program had spent more than a year building a rigorous process for production change. Sensitive deployments required review. Policies were documented. Exceptions were tracked. Approvals were recorded. By the standards most organizations use to judge progress, the program was working. Six months later, however, the company was breached; more importantly what the breach was allowed to become.

An investigation didn't reveal a reckless engineer, an ignored critical alert, or a glaring policy violation. It revealed something harder to admit: The organization had been looking carefully at each decision in isolation, while risk was accumulating across all of them. A deployment identity had gradually acquired broader authority through a sequence of legitimate, approved changes. Each step made sense locally. Together, they created a path the organization never intended to allow.

That's the moment we are in.

Over the past decade, security has matured. Organizations have invested heavily in cloud-native controls, identity systems, continuous monitoring, and automated enforcement. We can describe infrastructure at scale. We can audit access. We can detect drift. We can produce dashboards with a level of precision unimaginable not very long ago.

And yet, confidence has not grown in proportion to visibility.

Security teams still struggle to answer questions like: Which risks truly deserve immediate attention? Which issues can safely wait? Which seemingly reasonable changes are quietly expanding tomorrow's blast radius? Leaders are presented with precise findings but uncertain narratives. Engineers are asked to remediate individual issues without a clear explanation of how those issues would actually harm their systems.

This book begins from that tension.

It is written for security, platform, and engineering leaders working in environments where systems are continuously changing, authority is exercised through control planes, and automation increasingly outpaces human judgment. In such environments, the traditional mental model of security still produces many correct answers. However, it no longer produces enough useful ones.

The problem isn't that current security practices are naive. Most are internally coherent and technically defensible. The problem is that they were shaped in a world full of slower change, clearer boundaries, and more localized failure. In that world, evaluating the current state of a system was often enough to understand its risk. A snapshot in time could capture enough data to understand risk.

That world is gone.

Modern systems are dynamic, identity-mediated, API connected, and increasingly automated. Risk emerges less from single violations than from sequences of valid actions compounding over time. Temporary exceptions persist. Permissions accumulate. Trust relationships widen. Legitimate changes interact to create futures no one explicitly designed.

At the same time, organizations themselves have become more distributed. Teams optimize locally. Incentives shape behavior. Adoption cannot be assumed simply because a control is correct. Security outcomes increasingly depend on whether the system makes the safe path the natural path under real operational pressure.

Now, a further shift is underway. Autonomous and semi-autonomous systems are beginning to operate directly inside production environments.

These systems don't hesitate or apply contextual caution that human operators contribute informally. They exercise the authority they are given at speed and without the friction that used to mask deeper structural weaknesses.

This book argues that security must evolve accordingly.

Security can no longer be understood primarily as the evaluation of static posture. It must become the governance of trajectories: the shaping of what systems are allowed to become over time, through sequences, and by decisions. The central question is no longer only whether a component is correct in isolation. The question is which futures are possible, given the authority, interfaces, and workflows the system already contains.

I have spent close to two decades in cybersecurity across technical execution, consulting, product leadership, and organizational transformation. Over those years, I have watched the industry solve remarkable technical problems, while running into a more stubborn class of structural ones. This book is an attempt to use my systems-thinking mindset to clearly name that structural shift and to provide a practical way to respond to it.

The goal isn't to alarm or to endorse any specific set of products or tools. It is to *clarify the mental model and frameworks.*

The chapters that follow move from diagnosis to redesign to institutionalization. They examine why snapshot-based reasoning increasingly fails, why control planes have become the real arena in which risk is shaped, and what it means to redesign security as a discipline of structural constraint rather than downstream correction. Only once we see the problem correctly can we build systems that are safe at speed.

How to Read This Book

This book is designed to be used in layers. You can read it cover to cover, but you don't have to read it that way. The map below gives five paths through the material: one compressed path for readers with limited time, and four role-based paths. Each path tells you what to read first, what you can safely defer, and who else on your team should read the sections you skip.

If you have one hour: Read the Introduction, Chapter 2, and Chapter 4. That is the spine of the argument. The Introduction frames why the current model is failing. Chapter 2 names the model error: the snapshot fallacy. Chapter 4 introduces the replacement: governing what systems can become, not just what they are. If those three pieces land, the rest of the book becomes operating instruction for an idea you already hold.

If you are a CISO: Read Chapters 1–4, Chapters 10–11, and Appendices F–H. Chapters 1–3 give you the diagnosis. Chapter 4 gives you the replacement model. Chapters 10–11 translate that model into a leadership mandate, an operating loop, a scoreboard, an organizational design, and a transition plan. Appendices F–H give you the Decision Authority Register, the Big 5 high-entropy surfaces, and the Reachability Ratio toolkit. You can defer the implementation depth in Chapters 8–9, but your platform and security architecture leaders should read those chapters closely.

If you are a platform or security architect: Read Chapters 4, 8, and 9, then Appendices B–D. Chapter 4 gives you the framework. Chapter 8 shows how to build it: transition models, authority foundries, curated actions, sequence integrity, automatic collapse, and paved roads. Chapter 9 shows how to operate those constraints under pressure. Appendices B–D help with the most common design objections: why this is not just whitelisting, how to patch constraints after incidents, and why existing authorization models fail for agents. Read Chapter 10 if you need to present the work to executives or align closely with the CISO.

If you are a product, platform, or developer-experience leader: Read Chapter 6, Chapter 11, and Appendix I. Chapter 6 explains why security succeeds or fails inside organizations through defaults, friction, incentives, and propagation. Chapter 11 gives the transition playbook. Appendix I reframes the security product manager's job around adoption, speed dividend, paved-road usability, and deprecation of unsafe legacy paths. Read Chapter 8 if your team will help build the paved roads themselves.

If you are a senior individual contributor in a security function — staff vulnerability engineer, principal IAM architect, lead SRE on a platform team, cloud security lead, or AI safety engineer — read Chapters 1–4, Chapter 9, and the *Building the Institution* section of Chapter 10. Chapters 1–4 give you the diagnosis and reframe. Chapter 9 shows what operations look like inside decision space. The *Building the Institution* section names how the work inside security functions changes. Then choose the appendix closest to your domain: Appendix C for constraint patching, Appendix D for agent authorization, Appendix F for governance artifacts, and Appendices G–H for high-impact surfaces and reachability.

The chapters are written so each part can stand on its own, but the argument compounds. A reader who only reads Part I will understand why the current model is failing. A reader who reads Parts I and II will understand the replacement. A reader who reads all three parts will understand how to build it, operate it, and institutionalize it.

PART I

The Illusion of Control

Security today has more visibility than at any point in its history. Organizations can inventory infrastructure continuously, enumerate permissions, scan code and cloud configurations, review access, monitor endpoints, detect drift, and produce dashboards that summarize risk across thousands of systems. A security leader can walk into an executive meeting with charts, severity labels, control coverage, trend lines, and evidence that the program is working. This progress is real. It is also incomplete.

The problem is not that modern security lacks data. The problem is that much of the data describes the wrong unit of risk. It tells us what exists, what is exposed, what violates policy, and what has changed since the last scan. Those facts are important, but they do not always explain how risk will unfold, which paths are becoming reachable, or which ordinary decisions are quietly expanding future blast radius.

That is the illusion of control.

The illusion is not that security programs have no control. They do. They can block, detect, patch, review, approve, and respond. The illusion is that better visibility into the current state is the same as understanding future consequences. For a long time, that assumption worked well enough. Systems changed more slowly. Boundaries were easier to reason about. Human operators introduced friction. A permission, configuration, or network path often remained stable long enough for posture, security's current-state map, to carry real explanatory power. In that world, posture was a reasonable proxy for risk. That world has changed.

Chapter 1 begins with the gap practitioners feel every day: security findings can be precise yet insufficiently directional. A critical issue may look alarming in isolation while remaining operationally constrained. A medium-severity issue may appear less urgent when embedded within a workflow that can propagate authority across environments.

Chapter 2 names the deeper model error: the snapshot fallacy. A snapshot can describe what exists at a moment in time. It struggles to explain how the system is evolving through sequence, reuse, and accumulation. More frequent snapshots improve freshness, but they do not change the unit of reasoning.

Chapter 3 moves from the failure of snapshots to the place where risk is actually shaped: the control plane. Identity systems, CI/CD platforms, cloud APIs, Kubernetes control planes, SaaS administration layers, and agent tool interfaces do not merely administer systems. They author authority. They determine which futures become reachable.

The purpose of Part I is not to argue that current security practices are useless. They remain necessary. Scanning, patching, posture management, detection, access review, and incident response remains relevant. The problem is that they have often been asked to explain more than they can.

CHAPTER 1

Why Security Feels Precise but Fails in Practice

A posture review surfaces a critical finding: A service account has administrative access to a production database. This severity label might be technically defensible. In another part of the backlog sits a medium-severity finding: A broadly scoped role is attached to a continuous integration (CI) pipeline that deploys to twelve environments.

The critical finding gets escalated, while the medium-severity finding remains unaddressed. Both labels are correct by the logic that produced them. Yet, neither explains what would actually happen next.

The service account is clearly overpowered, but its real-world path to impact may be narrow and constrained. The CI role appears less urgent in isolation, yet it sits inside a workflow that can propagate change across environments, widen its reach, and become a bridge for lateral movement. One finding describes a condition. The other hints at a future.

Security teams encounter this gap constantly. They aren't drowning in backlog because they lack data. They are drowning because the systems producing that data often stop at description. They identify what is exposed, what is misconfigured, and what violates policy. What they struggle to explain, though, is which conditions will have the greatest impact, which can wait, and which small changes today quietly expand tomorrow's blast radius.

This is why modern security often feels precise but fails to inspire confidence. The findings aren't wrong; they are incomplete.

Precision Without Direction and Confidence

Over the last several years, security teams have gained extraordinary visibility into their environments. Infrastructure can be continuously inventoried. Permissions can be enumerated. Policies can be evaluated automatically. Dashboards summarize exposure with a level of detail that would've been difficult to imagine not long ago. This progress should've produced much greater confidence.

Instead, many teams feel a quieter and more uncomfortable tension. They can see more information than ever, yet still struggle to answer questions like: Which issue truly deserves immediate action? Which finding is mostly theoretical? Which approved change is quietly making the environment easier to compromise six months from now?

The problem isn't that modern security outputs lack precision. They are precise in a narrow sense. They can tell you that a certain permission exists, that a configuration deviates from policy, or that a resource is exposed. What they often cannot tell you is how risk will actually unfold.

A high-severity finding describes a condition. It rarely explains a trajectory. In practice, security teams are left to reconstruct, manually and repeatedly, the same unanswered questions: If this identity were to be compromised, what becomes possible next? Which combination of otherwise ordinary permissions would turn this exposure into real harm? Which seemingly minor change expands a system's future decision space in ways no one is seeing clearly?

It's because of this missing layer that security findings accumulate faster than confidence. Teams suppress alerts, postpone remediation, and make judgment calls that often look undisciplined from the outside. Their behavior, however, is usually rational. They are compensating for a deeper problem: The outputs in front of them are better at describing the current state than at explaining the consequences.

That isn't mainly a tooling problem. It is a contextual problem.

State Is Not Behavior

The limitation of the current state runs deeper than alert fatigue or imperfect prioritization. It reflects a more fundamental mismatch between what security systems usually describe and how modern failures actually happen. State is not behavior.

Modern incidents rarely unfold as a single dramatic violation. More often, they unfold as a chain of legitimate actions: An identity is enumerated, a role is assumed, a policy is adjusted, a trusted interface is invoked, a control-plane setting is changed, or a new path becomes available. Each individual step may be valid. Each may be authorized. Nothing in the sequence needs to appear alarming from the system's perspective.

That is precisely the problem, though.

Security systems that reason primarily about "state" can tell you that a permission exists. They can tell you whether a resource is public or whether a role is broad. What they struggle to represent is how a rational actor, human or automated, could use that permission, combine it with others, or exploit it as one step in a longer sequence of actions.

Return to the opening example. The critical service-account finding is easy to explain because it looks bad in isolation. The CI role is harder to rank because its danger lies less in what it is right now than in what it can connect, modify, and enable if used as part of a longer sequence. One finding fits a static model, while the other reveals its limitations.

This is why so many serious incidents seem obvious in hindsight and difficult to anticipate with confidence beforehand. In hindsight, the chain of events is visible. Before the fact, however, the model is usually asking the wrong question: What is true about the system right now? The more important question is: What becomes possible from here? A specified system state is useful but not sufficient to manage risk.

Distinguishing state from behavior is an important first step, but it isn't the end of the story. Behavior explains how a system arrived at a particular state. It does not explain why those actions occurred, who initiated them, or whether they should have been possible in the first place.

Answering these questions requires looking further upstream to the decisions that drive behavior and the authority that enables them.

The Illusion of Deterministic Severity

Because underlying uncertainty is very hard to model, security programs often compress complex conditional risk into deterministic labels such as critical, high, medium, or low. These labels are useful. They create order, help teams prioritize, and enable reporting. But they also create an illusion that risk can be classified with far more certainty than dynamic systems usually allow.

In modern environments, severity is rarely absolute. It is conditional.

A finding may be considered dangerous only if a specific workflow exists, a specific identity can reach it, a trust relationship has widened, or other small decisions have already accumulated around it. Another finding may look modest in isolation while quietly expanding what the environment can do next. The danger is not always in the immediate condition. It is often in the future that condition enables.

Deterministic severity flattens these differences. It rewards what is easiest to score in the present and often underrepresents what is most important about the future.

Practitioners notice this flattening quickly. Over time, they learn that some "critical" findings are operationally contained, while some "medium" findings are disproportionately consequential because they widen reach, connect boundaries, or increase the speed at which future damage can propagate. The result is predictable: The more confidently the system scores, the less confidence experienced operators place in that score.

What causes this to occur is an adaptation to a model that promises certainty where only conditionality exists.

The Assumption Beneath the Current Mental Model

The frustrations described so far aren't isolated flaws. Precision without confidence, state without behavior, and severity without consequence are all surface effects of a deeper assumption that has shaped much of modern security practice.

The assumption is simple: If we can accurately describe the current state of a system, we can adequately understand its risk.

That assumption wasn't careless. It emerged in environments where it was often good enough. Systems changed more slowly. Boundaries were clearer. Failure modes were more localized within a system. A firewall rule that was deemed unsafe on Monday was likely still unsafe on Thursday. A broad permission often represented a durable and legible exposure. The world held still long enough for a system's snapshot to carry explanatory power.

However, that's no longer the environment in which most organizations operate. Modern systems are continuously changing. Permissions are being granted to meet immediate needs. Policies are being adjusted under pressure. Resources constantly appear and disappear. Identities become the bridges through which trust, access, and operational dependency flow. Small changes that are individually reasonable can interact in ways that materially reshape what the system allows.

In such an environment, a system's snapshot is not necessarily inaccurate. It is simply being asked to explain too much. It can tell you what exists at a moment in time. It cannot, however, on its own, explain what the system is evolving into over time, through sequence, and through accumulation. It cannot show how a role became load-bearing through a series of isolated approvals. It cannot indicate how a narrow-trust relationship gradually evolved into a durable bridge across boundaries. It cannot display how a system's future possibilities have widened, although no single change looked dangerous enough to stop it from being made.

That assumption is known as the *snapshot fallacy*, which is the belief that the present state is sufficient enough to explain future risks. Snapshots remain useful. But once systems move fast enough, connect broadly enough, and delegate enough authority to automation, those snapshots then stop being an adequate model of risk or danger. At that point, they become partial descriptions mistaken for explanations of risk.

A Growing Mismatch

If the snapshot fallacy is so limiting, why did the model survive for so long? Because something compensated for it.

That something was friction.

For much of information technology (IT) security's history, human operators provided a buffer that the model never explicitly acknowledged. People acted more slowly. Changes required coordination. Approvals introduced delays. Experienced practitioners carried context in their heads. They knew which system was fragile, which permission existed for historical reasons, which exception was supposedly temporary, and which "technically acceptable" change made them uneasy for reasons that no dashboard could express.

When the security model missed something, judgment often compensated for the gap. That compensation did not always happen, and it did not happen reliably. However, it occurred often enough that the mismatch between the security model and reality remained tolerable.

Security teams still rely on this compensation constantly. These teams have an instinct to escalate issues when something "feels wrong," even when the report is clean. They may hesitate before making a technically compliant change that seems contextually dangerous. They hold informal knowledge that one identity, one workflow, or one integration matters more than the dashboard suggests. None of this information is captured cleanly in static evaluations. Human friction fills the space where the security model falls short.

That buffer is now eroding, however.

Automation compresses decision cycles. Infrastructure changes propagate faster than review processes can follow along. Integrations multiply the number of boundaries a single identity can cross. Systems increasingly provision access, modify state, and execute workflows on behalf of the human team. Autonomous or semi-autonomous systems remove the pause, hesitation, and contextual skepticism their human counterparts had that once served as an informal safety layer.

This situation reflects the operating reality of modern organizations. Speed, delegation, and automation aren't optional side effects. They are core features of how competitive digital systems now work.

But they remove the very buffer that made snapshot-based security tolerable.

When the distance between a decision and its consequence was measured in days or weeks, organizations had time to notice drift, question unusual changes, and correct course. As that measurable distance has collapsed in relation to the speed of an automated pipeline, the model's weakness has become direct operational exposure. It's that mismatch that is at the center of this book's discussion.

The issue isn't that security teams lack rigor. The issue is that many of the models they rely on were built for a world where a system's state was a better proxy for risk, human friction was a hidden compensator, and the system moved slowly enough that snapshots mattered more than its trajectories.

In modern environments, risk isn't best understood as a property of configuration alone. It is a property of decisions made over time.

If that's true, then no amount of additional precision applied only to the system's present state will fully solve the problem practitioners are already experiencing. Security doesn't simply need more visibility. It needs a better model of reality.

CHAPTER 2

The Snapshot Fallacy

A snapshot can be accurate and still mislead.

Modern security still assumes that if we can describe the present state of a system accurately enough, using its configurations, permissions, network relationships, and policy settings, we can then infer its risk. That assumption once worked well enough. In slower environments, the present state of a system often carries real explanatory power. A firewall rule, permission, or network path stayed stable long enough for the current state to serve as a useful proxy for future consequences.

That proxy is breaking. Cloud-native, SaaS-connected, identity-mediated, and increasingly autonomous systems do not merely sit in a state. They move through sequences of authority, trust, workflow, and control-plane change. A snapshot can tell us what exists now. It cannot, by itself, explain what the system is becoming, which paths have started to connect, or how ordinary changes are accumulating into future risk.

That is the *snapshot fallacy*: the belief that a sufficiently accurate description of the present state is enough to explain future risk. The problem is not that snapshots are wrong. The problem is that they are being asked to explain systems whose risk arises from change, accumulation, and sequence. This chapter names that model error.

When State Becomes a Poor Proxy for Risk

In stable environments, state and risk often move together.

When systems change slowly, access paths are narrow, and boundaries hold their shape over time, their configuration can tell you a great deal about how that system can fail. A firewall rule, a network path, or a permission boundary within that configuration represents a durable condition. If it is found to be unsafe today, it is likely to remain unsafe tomorrow. The system's environment holds still long enough for the snapshot to be explanatory.

Modern systems don't behave this way, however. Permissions are granted to solve immediate needs. Resources appear and disappear dynamically. Policies are adjusted under delivery pressure. Automation reuses identities across workflows that didn't exist when those identities were first created. Each change made may be legitimate, local, and temporary. Taken together, they reshape what the system allows.

A snapshot can tell you that a permission exists within a system. It usually cannot tell you why it exists, how it arrived there, whether it was introduced deliberately or inherited through reuse, or whether it will become dangerous only when combined with other otherwise acceptable conditions. It's this moment when a system's state begins to lose explanatory power.

Two systems can look similar in a posture report, while each carries very different risk profiles. One may be intentionally constrained. The other may reflect gradual erosion: permissions widened under pressure, exceptions preserved beyond their need, and trust relationships extended until boundaries that once mattered no longer offer real resistance. From the snapshot's perspective, those histories are invisible.

This is why so many security incidents appear obvious in hindsight. The relevant conditions were present, but their meaning was distributed across time and decision-making rather than being concentrated in a single misconfiguration.

The system's state did not lie. It simply didn't explain enough.

Why More Frequent Snapshots Don't Solve the Problem

Once this gap becomes visible, the most natural response is to increase frequency.

If risk emerges between scans, scan more often. If posture changes quickly, evaluate continuously. If drift is the problem, shorten the observation loop. These are sensible improvements, yet they aren't a solution. More frequent snapshots improve freshness. They don't change the unit of reasoning.

Each evaluation still asks some version of the same question: Is the system acceptable right now? What it doesn't ask, however, is a different class of questions:

- What new future did this change make reachable?
- Which previously separate paths are now connected?
- How does this decision interact with earlier decisions whose significance was invisible in isolation?

These questions are about transition, sequence, and accumulation.

This is why organizations often invest heavily in continuous posture and still feel uncertain. They can see more, yet they still cannot explain enough. Risk is being continuously observed and continuously fragmented.

It is worth acknowledging that attack-path analysis, graph modeling, and continuous posture tools are meaningful improvements over periodic scanning alone. Many of them show a system's present-state reachability far better than traditional reports do. But most still reason primarily from what the environment permits *now*. They are stronger at describing accessible paths than at explaining how authority accumulated, how a path became normal, or how a sequence of individually valid changes altered the system over time.

Watching the wrong unit more often doesn't make it the right unit to watch. The central failure in modern security isn't insufficient visibility. It is the use of a point-in-time description to explain a system's dynamic risk.

Risk as Accumulation, Not Violation

The missing unit is accumulation.

Much of security thinking is organized around violations. A configuration is compliant or noncompliant. A permission is acceptable or excessive. A control is present or missing. That framing once worked because, in more stable environments, discrete violations often did correlate closely with impact. In dynamic environments, however, that relationship weakens.

The most dangerous risks now often emerge not from a single obvious violation, but from the accumulation of multiple non-violations: decisions that are individually reasonable, individually reviewable, and individually easy to approve. A temporary exception lasts longer than expected. A narrowly scoped permission is reused in a broader context. A service account gains one more action, one more account, and one more dependency. Nothing flips cleanly from safe to unsafe. The system drifts.

Traditional security looks for a bad rule or a bad actor. Modern risk is an accumulation toward critical mass—the point where enough safe permissions align to form a trajectory that bypasses your defense-in-depth.

One safe permission rarely impacts much. Ten individually defensible permissions, concentrated in one identity or workflow, can produce something qualitatively different: not just more access, but enough accumulated reach to dissolve boundaries the organization believed were meaningful.

When a product team needs its analytics service to read customer records from a single DynamoDB table, a scoped IAM role is created with read access to that one table in one account. The request is then reviewed, approved, and deployed.

Say a few weeks later, the team needs billing data from a second account. A resource ARN is then added. Later, the service is expanded again; it now needs to consume an event stream for reporting. Another approval. Another policy edit. Then, a write path is added for derived metrics. Again, the change makes sense in context. Again, it is reviewed in isolation.

Notice what the reviewer actually sees each time. They don't see the role as an evolving system of authority. They see something different: One resource added, one action appended, one more immediate need. The question the workflow invites them to answer is not: What is this role becoming? It invites them to answer: Is this change acceptable?

But that question is too small.

By quarter's end, the role reads across multiple accounts and writes into a downstream metrics pipeline. What first began as narrow table-level access has evolved into a cross-boundary data path.

No single change was obviously reckless.

No single approval looked catastrophic.

The role may not even be "misconfigured" in the narrow technical sense.

But the organization has changed something deeper than simply a policy document.

The organization likely spent real effort creating account boundaries, segmentation, and data separation, so that compromise in one area wouldn't automatically spread to another. Through entirely legitimate and approved growth, a single identity has now begun to unify that blast radius. A compromise of the analytics service no longer requires an attacker to break multiple boundaries. The attacker inherits the role's accumulated scope, which now creates an authorized path across systems that the organization intended to keep meaningfully separate.

The role didn't bypass the boundary model. It became the bridge that made the boundary model less meaningful. This is the form of risk that snapshot-based security struggles to represent. A posture report may surface the role to the organization's attention. It may even flag it as having too many privileges. But the report still tends to present the issue as a list of permissions to clean up. What it doesn't naturally show is that the role has crossed a threshold: Enough individually acceptable grants have accumulated in one identity to create a new kind of future. That threshold is the real issue.

A single read permission may be safe.

A second read permission may still be safe.

A tenth read permission, concentrated in the same workflow, however, may create critical mass.

Not because any single permission granted is glaringly wrong, but because the accumulated authority now has enough reach to span accounts, unify trust boundaries, and turn a local compromise into a systemic pathway.

It's why violation-based models underweight the most consequential modern risks. They ask whether a specific rule is bad. They struggle to ask whether enough non-violations have accumulated in one place to make the system qualitatively more dangerous.

It's also why fixed severity labels so often disappoint experienced practitioners. Nothing in this story required a glaring "critical" misconfiguration. Each individual change might reasonably look to be of medium or even low risk in isolation. Yet the cumulative effect enlarges the organization's reachable future more than many isolated critical findings ever will. What mattered wasn't how bad each diff looked at the moment it was approved. What mattered was how much future decision space each diff quietly added.

One way of making this expansion more visible will be to ask how much reachable damage a role, workflow, or pathway now concentrates, which I'll cover later in the book. For now, the principle comes first: Risk grows not only when a permission is excessive on its own. It grows when authority compounds faster than the organization can see.

A trajectory-governed system wouldn't solve this issue by slowing down the approval process. It would solve it by changing the interface through which authority is expressed.

Instead of allowing a single role to grow through direct policy edits, the system would let the team request distinct capabilities like reading customer data, reading billing data, consuming an event stream, and writing derived metrics. The platform then would compile those needs into separate purpose-bound authorities with a clearer scope, separate blast radius, and better terminal conditions. The team could still move quickly. But the system would no longer make it easy to create a single bridging identity through accretion.

Risk is often created not by what becomes noncompliant, but by what becomes connected.

Decision Space and Trajectory

If risk is accumulation toward a critical mass, then decision space is the map of how that mass can move. Most security teams are trained to see the environment as an *inventory* of resources, identities, data stores, clusters, buckets, roles, and policies. This is a world of nouns.

But the system doesn't live only as an inventory. It also exists as a *graph of possible actions*:

what an identity can read,
what a pipeline can deploy,
what a role can assume,
what a workflow can trigger next,
what a control plane can modify.
This is a world of verbs.

A *snapshot* is very good at describing the inventory. It can tell you what nodes exist and what state they are in.

Decision space is broader: It's a graph of reachable states and actions allowed by the current configuration of authority, interfaces, workflow state, and trust relationships.

The distinction between the two is important, as systems are rarely breached or destabilized by inventory alone. They are compromised through movement—one action enabling the next, one trusted path leading to another, one identity carrying enough accumulated authority to traverse boundaries the organization thought were separate.

Reachability is part of the story but it's not enough. It asks a narrower, current-state question: What (inventory) can be accessed from here?

Decision space asks a broader architectural question: What meaningful next states and actions are reachable from here, by whom, under what authority, through which interfaces, and with what follow-on consequences?

Reachability tells you that a path exists. Decision space tells you what that path enables, how it composes with other paths, and which futures the system is now capable of reaching. If reachability is a map of what can be touched from here, decision space is a map of where the system can actually go next.

Viewed in this way, many environments have a decision space far larger than the set of states they actually intend to exercise. The reason isn't usually malice. It is accumulated authority: permissions granted "just in case," legacy trust edges kept for convenience, broad templates reused to avoid friction, and emergency pathways that never fully collapse.

Most of that authority may be idle most of the time, but it's still part of the decision graph. It still expands what the system is capable of doing next. It's why posture and risk often diverge. A snapshot can show a resource that looks fine in isolation. It is much weaker at showing how many paths now flow through it, or how much latent authority surrounds it. The inventory may appear stable, while the decision graph of reachable futures has expanded dramatically.

Trajectory is how a specific identity, workflow, or actor moves through that graph over time. Decision space defines what is reachable. Trajectory describes which reachable path is actually taken. That difference is the difference between capacity and motion.

A broad role may sit quietly for months.

An data pipeline may execute the same safe path thousands of times.

A trust boundary may remain dormant.

None of that changes the fact that the path exists.

When security failures occur, they are rarely failures of a single permitted action. They are failures of how valid actions compose into an unsafe trajectory: An identity enumerates, assumes, modifies, reads, deploys, or exports in a sequence the system technically allowed, but one that the organization never meant to authorize as a whole.

This is why present-state reasoning continues to underperform in dynamic environments. It scores conditions one at a time. Risk, however, is often embedded within the decision graph itself and can be found in:

which edges exist,

which identities connect too many nodes,

which paths cross too many boundaries,

and how much reachable future has been quietly accumulated in one place.

Traditional security asks: Is this resource configured correctly?

Trajectory governance asks the more revealing question: Is this path authorized?

A systemic breach is rarely the result of one misconfigured node in the environment. It is usually the result of an unintended trajectory through a decision space that was too large to reason about locally and too interconnected to remain safe by accident. Later in the book, this expansion of reachable future will become something we can measure more explicitly.

Where Risk Actually Takes Shape

The snapshot fallacy trains us to look for risk within infrastructure: the open port, the unencrypted volume, the broad role, or the exposed resource. Those configurations are usually the visible residue of earlier decisions, not the place where risk first comes into being.

Risk takes shape earlier in the process, at the moment authority is bound to an identity, a workflow, or a piece of code.

A permission is granted.

A token is minted.

A role is assumed.

A cross-account trust edge is established.

A pipeline is given the authority to deploy.

A temporary elevation is created for speed.

In a world of manual ticketing and human-speed change, we could afford to look at the results of a decision, because the result stayed the same for a longer period. In an autonomous world, however, risk is born at binding, the moment an automated system, an AI agent, or a CI/CD pipeline links a piece of code to a set of capabilities.

The deeper problem isn't that authority is sometimes too short-lived to be captured. It is that risk is already being created *before* the snapshot is taken. It's being created when a role is granted, a workflow is trusted, a pipeline is allowed to mint deployment power, or when a just-in-time elevation is attached to a path that was never meant to become normal. By the time monitoring detects the event, a more consequential choice has already been made: the system has been authorized to proceed along a trajectory that was never safe to enter.

This is why continuous monitoring isn't a corrective to the snapshot fallacy. It's a faster replay that improves observation. It doesn't govern the point where decision space is formed. The corrective action is moved upstream to control plane mechanisms that bind intent to authority and decide which transitions are reachable at all.

Control planes aren't merely management layers for provisioning and configuration. Control planes are the *real* attack surface. They are the systems that authorize movement, establish trust, and decide which transitions are reachable. In autonomous systems, their importance is amplified: the control plane defines the boundary between isolated action and cascading consequence.

A snapshot can describe authority after it exists; it cannot govern the moment authority is created. It can tell an organization that a role, token, trust edge, or deployment permission now exists. It cannot decide whether that authority should have originated, how far it may propagate, what future it makes reachable, or when it must collapse. By the time monitoring records the change, the more consequential decision may already have happened: the system has been authorized to move.

More frequent observation alone does not solve this problem. Snapshots remain useful descriptions of state, but they are the wrong primitive for governing dynamic risk. In dynamic systems, security must reason about where authority is authored, how intent becomes capability, and which futures become reachable as a result. To govern trajectories, security must move to the place where trajectories are formed. That place is the control plane.

CHAPTER 3

Control Planes Are the Real Attack Surface

The medium-severity CI pipeline role discussed in Chapter 1 mattered for a reason the severity label couldn't express. Risk wasn't especially alarming in a snapshot: A pipeline needed to deploy code, read configuration, and promote artifacts across environments. None of that sounds exotic. The risk wasn't that the role looked dramatic in isolation. The risk was that it lived within the control plane.

A control-plane foothold is different from a data-plane foothold. A vulnerability in one service may expose one system. A foothold in the control plane can change deployments, alter trust, widen permissions, and propagate a new state across many systems at once. It doesn't merely touch the environment. It helps decide what the environment becomes next.

Software exploits, exposed services, phishing, and application flaws are still important. They often determine entry. But in modern environments, the control plane often determines amplification and consequence. It is where local compromise becomes organizational risk.

Where Risk Is Shaped

Control planes are usually described as management layers. They provision infrastructure, define identities, attach policies, configure networks, and orchestrate deployments. That description is technically correct and strategically too small.

A control plane isn't just where a system is administered. It is where authority is authored.

In practical terms, this includes identity systems, continuous integration and continuous delivery/deployment (CI/CD) orchestrators, cloud management APIs, Kubernetes API servers, service-mesh policy layers, and the administrative surfaces of Software-as-a-Service (SaaS) platforms. Execution planes run workloads. Control planes determine which workloads exist, which identities they hold, how they connect, and how they can change.

Authority is the real lever in modern systems. The control plane defines who can create resources, who can connect them, who can deploy code, who can mint credentials, who can assume trust, and which changes become valid state. Those decisions are not simply bookkeeping. They determine the range of futures that the system can reach, which is why control planes are crucial, even when the initial point of compromise lies elsewhere. A web vulnerability might give an attacker code execution in one service. A stolen laptop might yield one developer credential. A leaked token might expose one repository. Those are entry events. Whether they remain local or cascade depends on whether they lead to a surface that can change authority, propagate configuration, or dissolve boundaries. That surface is usually the control plane.

The CI pipeline example makes this concrete. A pipeline role that can deploy to twelve environments isn't just a credential with broad access. It is a mechanism for rewriting runtime reality across those environments. If that role can promote artifacts, alter deployment definitions, inject configuration, or read the secrets needed to do those things, then compromising it doesn't simply expose one component. It gives an actor a way to shape what many components will do next. It's for this deeper reason that control planes deserve special treatment. They don't merely expose assets. They shape consequences.

The traditional security instinct is to ask: Where might an attacker get in? However, in modern systems, an equally important question is: Where could an actor, once inside, change an organization's future at scale? That is the control-plane question.

How Control-Plane Decisions Become Risk Trajectories

Control planes are important because they turn locally rational changes into durable organizational capability. There is a dangerous belief embedded in how organizations review change: If each step is reasonable, the resulting system must still be safe. That belief is the *myth of the reasonable step*. It sounds sensible, because each decision is usually reasonable when viewed in isolation:

A permission is added to remove friction.

A trust relationship is extended for consistency.

A pipeline is given a broader scope to speed delivery.

An automation is allowed to reuse an existing role because building a new one would take longer.

Each step is explainable. Often, each step is correct in the narrow context in which it is made. The problem is that control planes do not preserve the local reasonableness of each step. They compose those steps into durable authority, which is what organizations routinely fail to see. Reviewers are usually not asked: What is this system becoming? They are asked a smaller question: Is this change acceptable? The interface reinforces the mistake. A pull request shows a diff. A cloud console shows the current setting. A ticket shows an immediate need. Almost nothing in the default workflow shows the cumulative direction of the system. This is where the myth survives: in the gap between local judgment and global consequence.

Consider the CI pipeline role again. At first, the role is narrow. It deploys one service into one non-production environment. The scope is modest and easy to justify.

Say the team wants consistency, so the same pipeline pattern is extended to staging and production. Reuse is efficient. No one wants three different release mechanisms. Later, the pipeline needs read access to secrets, so it can fetch signing material and environment-specific configuration. That addition is approved, because the pipeline cannot do its job without it.

A few months later, rollback speed becomes important. The pipeline is allowed to update deployment definitions directly, rather than handing them off to a human operator. Then, a cross-account assumption path is added, so the same workflow can promote artifacts across multiple environments from a central build system. Every one of those changes is intelligible. Every change removes friction and can be defended as a practical improvement.

Now take a step back. What began as a delivery convenience has become a cross-environment control surface. The role can influence what code runs where, under which configuration, with access to which secrets, across which boundaries, and through a single workflow. No single diff says, "You are building an organizational blast-radius multiplier." That's precisely the point.

The most dangerous control-plane trajectories often do not look dangerous at the moment they are created. They look useful. They look efficient. They look like the reasonable next step. The failure isn't that people are irrational. The failure is that the system gives them no structural way to see when a sequence of rational decisions is converging into a form of authority that the organization never meant to create.

Audit logs, by themselves, don't solve the problem. They can tell you that a role was assumed, a policy changed, a deployment triggered, or a trust relationship updated. They describe events faithfully. What they do not naturally express is direction.

They don't tell you that authority is accumulating faster than it is collapsing. They don't tell you that a workflow is becoming a bridge across environments. They don't tell you that a formerly bounded mechanism is turning into infrastructure with systemic consequences.

Everything may be logged, but very little may be explained.

Decision space is the range of next actions the system makes reachable. *Trajectory* is how those reachable actions are actually composed over time.

Control-plane decisions don't merely operate inside the decision space. They author it. They widen or narrow the set of futures the environment can reach next. That is why control-plane risk isn't well described as a collection of bad settings. It is better described as the cumulative shaping of reachable futures through a sequence of valid moves.

Once that is clear, the weakness of exploit-centric thinking becomes easier to see. The exploit may determine entry. The control plane determines whether that foothold remains local or becomes systemic. A compromised service, credential, or agent becomes organizationally consequential only when it reaches a surface that can alter trust, expand authority, propagate change, or dissolve separation.

That is how local action becomes systemic risk.

The myth of the reasonable step is, therefore, not just a review failure. It is a governance failure. It is what happens when organizations evaluate control-plane changes one at a time, while the system silently composes them into a new kind of power.

It is also why autonomy raises the stakes so sharply. Once machine-driven systems begin operating inside the control plane, the sequence no longer unfolds at human speed. The system moves through those "reasonable steps" faster, more often, and without the hesitation that once provided an informal margin of safety.

Why Autonomy Collapses the Margin for Error

The myth of the reasonable step mattered even in human-paced systems. Under autonomy, it becomes far more dangerous.

A series of locally rational decisions was always capable of reshaping authority in ways no one intended. What made that tolerable for as long as it did was that human beings introduced an informal buffer into the system. People hesitated. They noticed when something felt broader than it needed to be. They slowed down when a shortcut seemed convenient but somehow wrong. They carried institutional memory about which boundaries were more fragile, more political, or more consequential than the interface itself revealed.

I call this the *stabilizing friction* of the human mind.

This stabilizing friction was rarely a formal control. It didn't live in policy, tooling, or enforcement logic. It lived in those moments where humans paused, had doubt, needed context, and passed judgment. It was the quiet resistance that kept some dangerous sequences from fully forming, even when the underlying system made them technically possible.

That friction did more work than most security models admit.

It helped older environments appear safer than they really were. It interrupted some bad sequences before they became incidents. It compensated, imperfectly but materially, for systems that were weak at reasoning about accumulation, sequence, and conditionality. In many organizations, the reason the old model held together as long as it did wasn't because it explained risk well. It was because people repeatedly patched over its blind spots in real time. Autonomy removes that patch.

A pipeline doesn't hesitate before using the authority embedded in its role. A remediation workflow doesn't second-guess whether the permission set it inherited is broader than the task requires. An agent doesn't carry social awareness about which boundary is load-bearing, which workflow is politically sensitive, or which shortcut only looked safe because a careful human used it sparingly. It executes.

The change is not only faster execution. It's a more consistent execution. A human might use broad authority reluctantly, occasionally, or only after reflection. A machine uses reachable authority as part of normal operation. What was once latent design debt becomes an executable pathway.

A permission that felt tolerable because a senior engineer exercised it carefully becomes a reusable machine-speed capability. A workflow that was once dependent on human coordination becomes a repeatable sequence. A trust relationship that seemed acceptable because it was rarely invoked becomes a standing propagation path once software begins to rely on it continuously. Autonomy collapses the margin for error. It narrows the distance between what the system is *currently allowed* to do and what the system *will eventually* do.

However, autonomy does more than accelerate familiar failure. It also introduces new categories of risk. Model-safety failures, prompt and context manipulation, unsafe tool composition, delegated authority exercised by probabilistic systems, and the coupling of reasoning engines to real control-plane power aren't just faster versions of older operational mistakes. They are new failure modes created by giving non-human systems the ability to interpret, decide, and act under uncertainty. That is the deeper break.

Appendix A develops this shift from first principles: Once systems interpret context, select actions, and pursue objectives over time, security must govern decision authority and trajectory, not merely state, output, or model behavior.

Older security models assumed that the actor was human, slow, and socially constrained. Even when authority was broader than it should've been, there were informal brakes: caution, habit, fear of consequences, or simple confusion. Agentic systems don't supply those brakes. They operate through prompts, retrieved context, and tool interfaces. They can misinterpret intent, compose actions unsafely, inherit authority that was never designed for them, and do all of it at machine speed. This is why authority becomes a first-order attack surface in autonomous environments.

The risk is no longer only that a system has access. The risk is that a probabilistic actor can now exercise that access in sequences no one explicitly enumerated, under conditions no human reviewer can track in real time. The question is no longer merely whether the action is authorized. It is whether that actor (human, automated, or agentic) should ever have been able to reach that action, in that sequence, with that degree of freedom, at all.

Autonomy is, therefore, both a reveal and a rupture. It reveals the authority design debt already embedded in the environment: the broad roles, inherited trust, standing permissions, and workflow shortcuts that human judgment once partially buffered. And it creates new operational risk by coupling non-deterministic reasoning systems to real authority over systems, data, and external effects. Such a combination is what makes the shift unavoidable.

A security model that depends on human review, hesitation, and contextual judgment to catch what structure failed to prevent is no longer resilient enough for the systems now being built. As execution speed rises and human pause disappears, safety can no longer depend on vigilance. It must depend on design.

If risk lives in evolving possibility rather than static posture, and if autonomy both removes the stabilizing friction that once masked bad authority design and introduces new machine-native failure modes of its own, then security can no longer remain primarily a practice of observing systems after the fact.

It must become a practice of shaping what systems are allowed to become before they move.

PART II

Redefining Security

Risk in modern systems is not best explained by a static state. It is shaped earlier, through how authority is created, widened, reused, and combined over time. Snapshots merely describe what exists. They struggle to explain what becomes possible next.

If that diagnosis is correct, then security doesn't simply need better visibility or faster detection. It needs to be redefined altogether. For nearly two decades, the security industry made an essential shift to *build security in.*

When software was the primary attack surface, security had to be engineered into its design, code, testing, and delivery. That shift gave us modern Application Security (AppSec). It improved security because, for a long time, behavior was still meaningfully bounded by artifacts. If you improved the architecture, the code, and the pipeline, you usually improved the security of the system.

That relationship is now weaker.

Modern systems are shaped not only by what is built, but by what is decided *after* the build ships. Permissions expand. Policies relax under pressure. Infrastructure changes continuously. Pipelines, platforms, and agents exercise authority that can materially alter what the system is able to do next. Even when the artifact is stable, the reachable future may not be.

This is the limit of *posture.*

Posture describes what the system looks like *now.* It doesn't adequately tell us what the system *can become.* That distinction is important because many of the most consequential failures in modern environments aren't the result of failures of a single artifact. They are failures of composition: a permission that becomes a pathway, a workflow that becomes a cross-boundary bridge, an agent that turns existing authority into a machine-speed attack surface.

Security, therefore, has to shift its center of gravity.

From artifacts to authority.

From state to possibility.

From evaluating conditions to shaping reachable futures.

This doesn't mean abandoning the axiom to "build security in." It means *extending it.*

If unsafe futures remain expressible, security is forced into a permanent cycle of detection, correction, and cleanup. If unsafe futures are removed from the system's grammar, speed and safety no longer pull in opposite directions.

It's the shift from posture to trajectory governance.

CHAPTER 4

Trajectory Governance

The first era of software security taught us to govern what we build. The next era must teach us to govern what our systems are allowed to become.

Trajectory governance is the practice of shaping which sequences of action are possible within a system. It is concerned not only with whether an individual action is allowed, but with whether a dangerous path should be reachable at all.

A deployment runs. A role is assumed. A token is minted. A policy changes. A workflow crosses a boundary. A supplier tool interacts with an internal system. An agent calls a tool and updates the state. Each action may be valid in isolation. The danger often appears in the sequence.

Trajectory governance asks whether those valid steps can combine into an unsafe future. This doesn't mean abandoning posture management. But posture describes what the environment looks like now. It doesn't, by itself, explain what the environment can become. Trajectory governance extends the focus from state to possibility.

It asks what systems can reach, what authority can widen, what workflows can influence, what trust can propagate, and what paths become available after each step succeeds. Before an organization can govern those paths, it must first understand the space of possibilities available to the system.

That space is the *decision space*: the set of meaningful actions, transitions, and future states available to a system given its current authority, workflows, interfaces, and trust relationships.

The Decision-Space Gap

Most security programs are built to understand posture.

They inventory assets. They scan for vulnerabilities. They monitor endpoints. They review configurations. They track cloud resources. They classify alerts. They report on compliance. They measure whether known systems match expected states. All of this is necessary. An organization cannot secure what it cannot see. But posture has a limitation. It describes what exists. It doesn't necessarily explain what can happen next.

Modern risk often emerges from connection, not isolation. A vulnerability is not only a flaw in a system. It may also be a path into another system. A permission is not only an assigned right. It may also be a bridge into a broader authority plane. A workflow is not only a sequence of approved steps. It may also be a mechanism for transferring trust from one context to another. This is the organization's decision space.

Decision space is the set of meaningful moves available inside an environment. It includes what users can do, what software can do, what systems can reach, what workflows can influence, what authority can be inherited, and what paths become available after an action succeeds.

Security programs often do a poor job of understanding this space. They may know that a server is vulnerable. They may know that a service account is privileged. They may know that a CI/CD pipeline can deploy to production. They may know that a SaaS integration has access to sensitive data. And they may know that a cloud role can assume another role.

But these facts are often analyzed separately. The more important question is how they compose.

Can the vulnerable server reach the privileged service account? Can the CI/CD pipeline alter a trusted release path? Can the SaaS integration influence identity or policy? Can the cloud role assume another role that changes its effective authority? And can a compromised workload inherit credentials that allow it to move into a more sensitive environment? These are not posture questions. They are decision-space questions.

This is why many security programs are rich in findings but poor in context. They know where weakness exists, but not always where weakness can lead. They know what is misconfigured, but not always what the misconfiguration makes possible. They know which assets are exposed, but not always which exposed assets can expand authority.

The risk is not only in the object. The risk is in the path.

A critical vulnerability on an isolated system may be urgent. But a medium-severity vulnerability on a deployment runner, identity connector, secrets broker, or cloud control-plane component may be more consequential because of what compromise makes reachable afterward.

The same is true for permissions. A broad permission on a low-impact system may be manageable. A narrower permission in the wrong workflow may become dangerous if it allows a system to influence deployment, identity, policy, or secrets.

Trajectory governance therefore depends on a question that most posture programs are not built to answer: *If this system falls, what becomes possible?* The question is simple. Answering it is not.

It requires understanding how the environment actually operates: how code reaches production, how identities receive permission, how secrets are accessed, how supplier tools interact with internal systems, how exceptions accumulate, and where authority can widen.

Most organizations do not lack security data. They lack a connected security understanding. The relevant data exists in many places: vulnerability scanners, identity providers, endpoint tools, cloud platforms, CI/CD systems, secrets managers, SaaS applications, ticketing systems, incident records, access reviews, and business workflows. Each source provides a partial view. But trajectory governance depends on understanding how those partial views combine into reachable paths.

This is where applied security data science becomes essential. Whether that capability sits in security engineering, risk analytics, platform security, or a dedicated data science function is less relevant than the work itself: turning fragmented operational data into insight about reachability, authority, dependency, and consequence.

Applied security data science asks different questions from a traditional reporting function. Not only: How many critical vulnerabilities exist? But also: Which vulnerable systems also hold privileged access? Not only: How many privileged accounts do we have? But also: Which identities accumulate permissions over time? Not only: Which systems failed policy? But also: Which workflows can push code, policy, or configuration into production? Not only: Which vendors are connected? But also: Which supplier-controlled tools have privileged reach inside the environment? These questions reveal where consequential choices happen.

They also reveal where security, platform, and product teams need to change the system itself. If teams repeatedly take risky paths, the answer is not only to warn them more often. It is to understand why those paths are available, why they are convenient, and what safer default path should exist instead. This work cannot be done by scanners alone.

It requires analysis, product judgment, and operational knowledge. Security has to understand not only what is wrong, but how work actually gets done. Otherwise, it will keep treating symptoms as isolated findings while the same unsafe paths remain available.

Most organizations inventory infrastructure better than they understand authority. That gap becomes more dangerous as environments become more software-defined. In cloud systems, authority can be created through roles, tokens, policies, service accounts, federation, deployment systems, and automation. In SaaS environments, authority can propagate through integrations, delegated permissions, workflow automation, and administrative APIs. In CI/CD systems, authority can move from code to artifact to deployment to production. In AI-enabled systems, authority may be delegated to agents that can reason, act, call tools, update state, and influence future decisions.

In each case, the question is not only whether the current state is approved. The question is what future states the current design allows.

A posture view can tell an organization that a role exists. A decision-space view asks what that role can become part of. A posture view can tell an organization that a pipeline has passed. A decision-space view asks whether that pipeline can alter trust. A posture view can tell an organization that an integration is authorized. A decision-space view asks what authority that integration can propagate if compromised.

Trajectory governance begins when the organization makes this space visible enough to reason about. This doesn't mean every possible path must be mapped perfectly. Perfect knowledge is not the goal. Security teams do not need a complete simulation of the enterprise before they can act. They need enough understanding to identify where consequential choices happen: where authority is minted, where trust is propagated, where access expands, where deployment occurs, where exceptions accumulate, and where local compromise can become a systemic consequence.

Without this understanding, security teams will continue to prioritize the wrong things. They will treat all critical vulnerabilities as equally urgent, even when some are isolated and others sit next to authority. They will approve exceptions without understanding the future paths those exceptions preserve. They will deploy controls where visibility is easiest rather than where intervention matters most. They will mistake local compliance for systemic safety.

The decision-space gap is therefore not only a visibility problem. It is a governance problem. Security cannot govern trajectories it doesn't understand.

The first task of trajectory governance is to make the organization's possible moves visible enough to reason about. Once that happens, familiar security questions begin to change.

Instead of asking only: Is this system vulnerable? The organization can ask: If this system falls, what becomes possible? Instead of asking only: Is this access approved? It can ask: What future authority does this access enable? Instead of asking only: Did this workflow follow the process? It can ask: Does this process allow trust to be influenced by the wrong actor?

These questions move security from observing posture to governing possibility.

They prepare the organization for the next step: deciding which futures should remain reachable, which should be constrained, and which should be made structurally impossible.

The Limits of Behavioral Governance

Once the decision space becomes visible, another limitation emerges.

Behavior provides more context than posture. A posture report may show that a permission exists. Behavioral evidence can show that the permission was exercised. A posture review may identify a broad role. Behavioral data can reveal which systems used it and when. It helps explain how a system arrived at its current state.

It doesn't fully explain why those actions were possible in the first place. Behavioral governance attempts to manage risk by observing actions as they occur. Detection systems, monitoring platforms, and anomaly engines look for signals that suggest a trajectory is becoming unsafe. These capabilities remain important. Organizations still need to investigate incidents, understand operational reality, and respond when systems behave unexpectedly.

The limitation is that behavioral governance operates after movement has begun. By the time a behavioral system evaluates an action, the system has already entered the trajectory. Behavioral governance operates entirely downstream, at the execution plane. This means an unsafe trajectory must remain possible before the system can catch it. If an environment allows unsafe trajectories to remain reachable, the security program is trapped in a permanent, resource-intensive cycle of monitoring, ticketing, and clean-up.

This is why security cannot stop at behavior. State describes what exists. Behavior explains how the system moved. Neither fully explains why a particular path was reachable. These answers lie further upstream, in the design of authority, workflows, interfaces, and trust relationships. They lie in the possibilities the system makes available.

Governing Possibility, Not Behavior

Once the decision space becomes visible, the purpose of security governance shifts. The question is no longer only: Is this action allowed? The question becomes: Should this future be reachable? That is the shift from governing behavior to governing possibility.

Behavior is what a system does in a given moment. A deployment runs. A role is assumed. A policy is changed. A token is minted. A resource is exposed. A workflow crosses a boundary. Possibility is different. Possibility is the set of actions, transitions, and sequences the system can express at all. It is shaped by architecture, interfaces, permissions, workflows, authority boundaries, supplier integrations, and control planes.

Many dangerous outcomes do not look dangerous at the level of individual action. A role assumption may be approved. A deployment may follow the documented process. A service account may hold a permission for a legitimate reason. A supplier integration may be authorized. A vulnerability exception may be accepted. A workflow may pass every required check. Each decision can appear reasonable in isolation. The danger appears when those decisions compose.

A role assumption becomes a path into another environment. A deployment workflow becomes a mechanism for propagating trust. A service account becomes a standing source of authority. A supplier integration becomes a privileged bridge. A vulnerability exception becomes a decision to preserve a reachable attack path. The individual action may be valid. The sequence may still be unsafe. Trajectory governance is concerned with that sequence.

It asks what the system can become after each approved step succeeds. It asks what new choices become available, what authority widens, what trust propagates, and what paths remain open. This is why trajectory governance is not the same as simply enforcing stricter rules. A rule says whether an action is allowed. A trajectory asks what that action makes possible. To some, trajectory governance can sound, at first, like allowlisting. But allowlisting usually governs whether a specific action, actor, resource, or condition is permitted. Trajectory governance governs whether permitted actions can compose into unsafe sequences. The difference is not merely what is allowed. It is what becomes reachable after something is allowed. For a detailed distinction, see Appendix B.

This distinction is especially clear in vulnerability management. A traditional vulnerability management program asks: How severe is this vulnerability? Severity, exploitability, exposure, and remediation time all matter. But they are not enough.

Trajectory governance adds another question: *If this system falls, what can an attacker do next?*

That question changes the meaning of vulnerability management. A critical vulnerability on an isolated internal system may be urgent. It may require fast remediation. But if the system has limited connectivity, no meaningful privileges, and no path into identity, secrets, deployment, or production authority, the consequence may remain local.

A medium-severity vulnerability on a CI/CD runner, identity connector, secrets broker, cloud control-plane component, or privileged SaaS integration may be more consequential. Not because the vulnerability is technically worse. Because the trajectory is worse. If that system falls, the attacker may be able to inherit deployment authority, access secrets, modify release artifacts, assume roles, alter policy, influence identity, or move into a more sensitive environment. The vulnerability matters because of what it can become.

This reframes vulnerability management. The goal is not only to reduce the number of findings. It is to reduce the number of findings that sit on paths to broader authority. A posture view asks: Which vulnerabilities are critical? A trajectory view asks: Which vulnerabilities can become control?
That is a different prioritization model.

It doesn't discard severity scores. It puts them in context. A critical vulnerability, an internet-facing exploit, and a known exploit remain important. But trajectory governance adds the missing dimension: consequence after compromise. How far can the attacker go afterward? That question pulls vulnerability management out of the scanner and into the architecture of the environment. It requires understanding reachability, identity, deployment paths, secrets' access, supplier trust, and authority boundaries.

It also changes how exceptions are understood. A patch delay on an isolated system is one kind of decision. A patch delay on a deployment system is another. A patch delay on an identity broker is another issue. In each case, the organization is not merely accepting a vulnerability. It is deciding which future paths remain open while that vulnerability exists.

The same is true for compensating controls. A web application firewall may reduce exploitability. Endpoint detection may increase visibility. Network segmentation may slow lateral movement. These controls may be useful and necessary. But trajectory governance asks a sharper question: Does the unsafe path still exist?

If the vulnerable system can still reach privileged credentials, influence deployment, modify identity, access secrets, or bridge environments, then the organization may have reduced observable risk while preserving structural risk. The trajectory remains reachable. This is where consequential decisions begin to surface.

Many of the most important decisions in security are not framed as strategic decisions at all. They appear as routine operational choices: Delay this patch. Approve this exception. Keep this integration active. Allow this service account to retain access. Permit this pipeline to deploy across environments. Let this supplier tool remain connected. Extend this temporary access. Accept this compensating control.

Each decision may be practical. Each may be justified. Security cannot pretend that every risk can be removed immediately or that every system can be redesigned at once. But trajectory governance requires these decisions to be recognized for what they are. They are decisions about reachable futures.

A vulnerability exception is not only a delay in remediation. It may be a decision to preserve a path from weakness to authority. A supplier integration approval is not only a vendor decision. It may be a decision to permit an external system to participate in internal authority flows. A long-lived credential is not only an access-management issue. It may be a decision to allow authority to persist beyond the task that required it. A broad deployment role is not only an operational convenience. It may be a decision to let one workflow span boundaries that should remain separate. These decisions belong to governance because they shape what the system can become.

Governing possibility doesn't mean predicting every attack. It doesn't mean modeling every path perfectly. It means identifying the places where ordinary operational choices materially change future reachability. It asks where authority can widen, where trust can propagate, where local compromise can become a systemic consequence, and where an approved action can open an unsafe sequence. These are the questions that trajectory governance brings into everyday security work.

Governing possibility means making those future effects explicit. It means asking not only whether something is acceptable now, but what it allows later.

It means recognizing that the most important security decision may have occurred before the alert, before the exploit, before the incident, and before the dashboard turned red. The decision may have been to leave the path open.

Trajectory governance brings that decision into view. It gives security a different object of governance: not only the action, not only the behavior, not only the present state, but the reachable future.

Controls That Shape Trajectories

Once security begins to govern possibility, controls have to be evaluated differently. The important question is not only whether a control exists or whether it works in a narrow technical sense. The more important question is where the control acts on the trajectory.

Some controls act after harm has begun. They contain damage, revoke access, isolate systems, restore service, and help the organization survive failure. These controls are necessary, but they do not change whether the unsafe path was reachable in the first place. Other controls act while a risky transition is being attempted. They inspect context, apply policy, request approval, narrow access, or block an action before it completes. These controls can reduce exposure and prevent many unsafe moves from succeeding. But they still depend on the system expressing the risky move first.

A smaller and more powerful class of controls acts earlier than both. These controls change the system so that certain unsafe paths are not available as normal options. They do not merely catch a bad move. They remove the move from the system's reachable decision space. Trajectory governance depends on understanding the difference.

Security programs often use the same words for all three categories: controls, policies, checks, guardrails, approvals, and enforcement. That shared language hides an important distinction. A control that alerts after a bad deployment, a control that blocks the deployment at runtime, and a control that makes the bad deployment structurally impossible are not doing the same kind of work.

They act at different points in the trajectory. They govern different things.

Level 3: Response Controls

Level 3 controls operate when harm is already occurring, or when the unsafe trajectory has already produced visible risk. This is the familiar world of incident response and containment. Security isolates systems, revokes credentials, rolls back deployments, disables accounts, restores from backup, rotates secrets, blocks traffic, or limits blast radius.

In the CI/CD example from Part I, Level 3 means the pipeline has already pushed a bad artifact, propagated a harmful configuration, or used broad deployment authority in a way that creates production impact. Security responds by revoking access, rolling back the deployment, isolating affected systems, and restoring service. This work is essential.

No security program can ignore response. Systems fail. Controls miss things. New attack paths appear. People make mistakes. Vendors are compromised. Response controls help the organization survive those failures. But response controls govern impact. They do not govern whether the unsafe trajectory was reachable in the first place.

A rollback may undo a bad deployment, but it doesn't explain why the pipeline could create that deployment. Credential revocation may stop active misuse, but it doesn't explain why the credential was broad, long-lived, or reachable. Isolation may contain a compromised workload, but it doesn't explain why that workload had a path into more sensitive systems.

Level 3 is necessary. But if the same class of failure keeps returning, the response is not enough. The organization is surviving the trajectory, not governing it.

Level 2: Guardrails

Level 2 controls operate when unsafe behavior is allowed to appear briefly, then corrected. This is where much of modern security tooling lives. A scanner detects a bad configuration. A cloud security tool flags an exposed resource. A policy engine blocks a deployment. A runtime guardrail reverts a risky change. A ticket is opened. An alert is created. A workflow is interrupted before the condition becomes worse.

In the deployment example from earlier, the pipeline may still be able to push a bad artifact or apply an unsafe configuration. A scanner, policy engine, or deployment guardrail detects the condition and fails the deployment, rolls it back, or opens an alert. This is better than discovering the issue weeks later in production.

Guardrails reduce dwell time. They shorten feedback loops. They make unsafe behavior more visible. They help teams catch mistakes before those mistakes become incidents. But guardrails still preserve the unsafe possibility within the system. The system still knows how to express the bad move. Security is trying to catch that move quickly enough.

A guardrail that blocks an overly permissive cloud policy is useful. But the workflow still allowed someone or something to request that policy. A deployment gate that rejects an unverified artifact is useful. But the pipeline still allowed the unverified artifact to arrive at the gate. A scanner that catches a secret in code is useful. But the development path still allowed the secret to be committed.

Level 2 controls govern correction. They do not fully govern possibility. They are especially valuable when systems are changing quickly, when the organization is still learning the risk pattern, or when full structural prevention is not yet practical. But they should not become the permanent answer for risks that are already well understood.

If a risky path is known, recurring, and high-impact, the goal should be to move it out of Level 2 over time. It should become a design problem.

Level 1: Structural Constraints

Level 1 controls operate at the system-design level. They change the environment so that an unsafe trajectory is not reachable through the normal path of work. This is the strongest form of trajectory governance.

At Level 1, constraints are embedded into architecture, interfaces, authority boundaries, workflow states, and valid transitions. The system doesn't merely detect or reject the bad move. It doesn't offer the bad move as a normal option. In the CI/CD example, the pipeline cannot deploy to production without a validated artifact. It cannot promote across environments through one broad standing role. It cannot widen its own authority. It cannot read arbitrary secrets because that would be convenient. Production deployment authority is minted only for a specific service, artifact, environment, and workflow state, and that authority collapses when the workflow ends.

The unsafe sequence is not detected faster. It is absent.

There is nothing to alert on, nothing to roll back, and nothing to remediate for that class of failure because the system doesn't offer the unsafe trajectory as a valid path. This is the level at which security begins to govern possibilities directly.

Level 1 also changes how security is experienced. At Levels 2 and 3, security often feels like an interruption. Work is blocked. Access is revoked. A deployment is reversed. A team has to stop, explain, remediate, or escalate.

At Level 1, the experience can be different. The safe path is built into the system. Developers and operators do not have to pause at every step to decide whether a choice is acceptable. Automation doesn't have to hold broad authority and hope a later check catches misuse. The environment has already been shaped so that ordinary work proceeds inside safe boundaries. This is why structural constraints can enable speed.

They do not slow the system down by adding another checkpoint. They reduce the number of checkpoints the system needs. They remove unsafe variation from the supported path so that useful work can move faster with less review, less rollback, and less remediation.

Level 1 governs decision space. It changes what futures the system can form.

The following table summarizes these three levels.

Table 4.1: Three control levels with what each governs and how

Level	What It Governs	How It Works	Best Use
Level 1: Structural Constraints	Decision space/ possibility	Unsafe trajectories are unreachable by design. Constraints are embedded into architecture, interfaces, authority boundaries, and valid transitions.	Known, recurring, high-impact workflows where safety must hold at speed and scale.
Level 2: Guardrails	Behavior correction	Unsafe behavior is permitted briefly, then detected, blocked, reverted, or remediated.	Emerging risks, evolving systems, and cases where full structural prevention isn't yet practical.

Level	What It Governs	How It Works	Best Use
Level 3: Response Controls	Impact containment	Harm is already occurring. Controls focus on containment, rollback, isolation, and recovery.	Unknown failures, active incidents, and resilience when prevention fails.

The goal isn't to eliminate Levels 2 and 3. Mature security programs *need* all three levels. Response controls are essential because no system is perfect. Guardrails are essential because environments evolve, constraints degrade, and new failure modes appear before they are fully understood. Levels 2 and 3 provide resilience, recovery, and learning. But they should not be the primary operating model for known, recurring, high-impact workflows.

There is a control-level mismatch in many security programs. Organizations invest heavily in detection and response, tune alerts, automate remediation, and add more policy checks, yet still feel exposed. They are working hard at Levels 2 and 3 while risk continues to be introduced at Level 1, where authority, interfaces, workflows, and transitions are defined.

Trajectory governance pushes repeatable failure classes upward. A pattern that first appears as an incident may begin at Level 3. Once understood, it may move into Level 2 as a guardrail or policy check. Once it becomes predictable, recurring, and high-impact, it should move into Level 1 as a structural constraint.

A dangerous trajectory should not remain an alert forever. It should not remain a recurring exception forever. It should not remain a rollback procedure forever. If the organization understands the path and can prevent it structurally, then the path should be redesigned.

Level 3 helps the organization survive failure.

Level 2 helps the organization correct unsafe behavior.

Level 1 helps the organization move fast and safely because unsafe trajectories are no longer reachable by default.

This is why trajectory governance changes the economics of security. A response-heavy model scales with the number of incidents. A guardrail-heavy model scales with the number of risky actions. A structural model scales with the design of the system itself.

When unsafe paths are removed from the decision space, the organization no longer has to detect, debate, approve, reverse, or recover from that class of failure each time it appears. The control becomes part of the environment. Safety is no longer dependent on perfect judgment at every moment of execution. At Level 1, security becomes most enabling.

At Levels 2 and 3, security often feels like control over the business. Actions are blocked. Access is revoked. Work is interrupted. Teams slow down because they must account for possible corrections, rollbacks, reviews, and escalations.

At Level 1, security feels less like control and more like a better operating path. Unsafe options are not presented as normal choices. Developers are not asked to rediscover the same failure modes. Automation doesn't need broad-standing authority. Operators do not have to choose between speed and safety every time they act. The system has already made some unsafe futures unreachable. That difference matters in three ways.

Freedom of Thought vs. Freedom of Impact

The most persistent objection to structural constraints is that they limit creativity. This objection rests on a false equivalence that limiting outcomes also limits thinking. In reality, structural constraints don't govern thought; they govern impact.

An engineer is free to imagine any architecture, explore any idea, and test any hypothesis. That freedom of thought is essential to innovation. Structural constraints don't interfere with that freedom of thought. What they manage is the blast radius of what those ideas can do when they interact with real systems with data, users, and authority.

The constraint is not on imagination. The constraint is on unsafe impact.

A useful analogy is a canvas where an artist can paint anything they want. The canvas is a constraint, but it doesn't kill creativity. It defines the surface on which creativity becomes meaningful, durable, and shareable. Structural constraints work the same way. They don't restrict imagination; they define the surface on which imagination can safely act. Without constraints, every idea carries more operational burden. Every new workflow needs review. Every exception needs debate. Every risky path needs a later control. The environment becomes harder to move through because too much is possible.

Good constraints narrow unsafe impact so that useful creativity can move faster. They do not make engineers less creative. They make the consequences of creativity safer.

Late Blocking vs. Informative Design

Structural constraints also change how security is experienced. When controls appear late, they feel punitive. Work is blocked after an effort has already been invested. A decision is reversed after intent has formed. Even when correct, late-stage controls feel like a judgment call made by an external authority. It is the experience that most people associate with security.

Structural constraints are encountered early, in the architecture, platform, interface, and supported path itself. They surface as information, not as rejection. The system simply offers a limited set of valid options, and unsafe variants are never presented as choices.

When unsafe options are absent, developers don't feel constrained; they feel guided. Decision fatigue is reduced. Cognitive load drops. Security assumptions are encoded into the environment itself, rather than enforced through interruption.

This is why well-designed constraints are experienced not as friction, but as features. Structural constraints make the safe path easier to see, to choose, and to trust.

Scaling Beyond Human Judgment

The most important property of structural constraints is how they scale.

Behavior-centric security depends on observation, interpretation, and intervention. As systems accelerate, the cost of that model rises with activity: more actions, more alerts, more reviews, more analysts, and more automation to monitor more automation. Human judgment remains the hidden bottleneck in the systems.

Structural constraints scale differently.

When unsafe states are structurally unreachable, the cost of safety doesn't increase with the number of actions taken. Whether a system supports 10 developers or 10,000, whether it executes hundreds of changes a day or millions, the same constraints apply. Safety is enforced by design, not vigilance.

That isn't a minor efficiency gain. It is a different operating model.

Structural constraints make security less dependent on perfect judgment at runtime. They allow humans, automation, and eventually autonomous systems to operate at speed because the environment itself has already removed certain dangerous variants from the reachable decision space. An autonomous system doesn't need a policy manual in the same way a human does. It needs a bounded decision space. It needs hard limits on what tools it can call, what authority it can inherit, what state it can modify, what data it can reach, and what actions require brokered approval.

It doesn't merely make systems safer. It makes speed safer.

A system governed mainly by Levels 2 and 3 security controls must constantly slow down, inspect, correct, and then recover. A system governed increasingly through Level 1 can move faster because fewer unsafe paths are available in the first place.

That is why constraint design isn't just a security primitive. It is an operating primitive for high-speed systems.

Structural Constraints as the True Shift Left

A common response I hear is: "We already do this. We've shifted left." We've spent years moving security earlier in the development lifecycle. We stopped waiting until production to reason about risk. We moved checks into design, code review, continuous integration/continuous delivery/deployment (CI/CD), and deployment. It was real progress.

In many organizations, however, the shift left changed when security intervened more than it governed. Shift left improved chronology. It did not, by itself, change the plane of governance.

A developer writes code, commits it, and is told by a scanner, pipeline check, or policy engine that something is wrong. The feedback arrives earlier than it would have in production, which is useful. But the unsafe variant was still expressible. The developer could still write it, submit it, and wait for security to reject it (i.e., earlier enforcement).

Constraint design makes a different move. It changes the environment in which the work is performed, so that unsafe variants aren't even options in the first place. Developers work through higher-level abstractions, supported workflows, and curated actions that already encode security-critical requirements.

The "no" doesn't simply arrive earlier. The unsafe choice disappears from the normal work trajectories. This is why structural constraints are the true completion of shifting left. The goal was never merely to move pain closer to the developer. The goal was to move security into the system deeply enough so that the developer no longer has to rediscover the same failure modes at all.

The change required is not primarily chronological but ontological. This shift also changes the economics. A conventional shift left still depends on a scan-and-remediate cycle. If unsafe states are expressible, they must be detected. If they are detected, they must be triaged. If they are triaged, they must be ticketed and fixed. As systems grow, that cycle grows with them, adding more scans, more dashboards, more exceptions, and more queues.

Structural constraints break the cycle for the classes of failure they remove. When a failure mode is made structurally impossible, there is nothing to scan for in that form, and nothing to remediate after the fact. The organization stops maintaining an ever-expanding list of forbidden states and starts designing a smaller and more intentional universe of valid paths.

Conventional shift left asks how early the organization can catch the problem. Trajectory governance asks why the problem is expressible at all.

When security operates at that level, the most important security artifact is no longer only a scanner, dashboard, or policy report. It is the set of abstractions through which developers, operators, platforms, and agents are allowed to act.

The paved road becomes the control. Not because it reminds people about what to do, but because it changes what normal work can express. (more on this in Chapter 8.) In this sense, structural constraints aren't an alternative to shifting left. They are the completion of shift left. Security doesn't need to happen only earlier. It needs to happen differently.

If unsafe futures are shaped by how authority is created, widened, and reused, then security must pay special attention to the systems that mint authority in the first place.

CHAPTER 5

Authority, Responsibility, and Distribution of Risk

No one has to be careless for an organization to become unsafe. A team grants permissions to unblock delivery. Another team reuses an existing role because it already works. An exception is approved for launch and yet is never fully removed. A shared workflow expands because every new consumer makes it more valuable. Each of these decisions is easy to defend. Each has an owner at the moment it is made. Each appears to solve a real problem.

Then time passes. The organization's team reorganizes, and its service gains new consumers. The workflow then becomes infrastructure, and the exception becomes the new normal. The role becomes load-bearing. The people who first understood the organization's original intent move on, but the authority they had remains exactly where the control plane left it. That is the organizational failure mode on which this chapter is focused.

Security must shape what systems are allowed to become. But organizations often produce the opposite: They let authority persist, expand, and compose long after responsibility has become diffuse.

Authority scales mechanically. Responsibility scales socially.

Authority is encoded in systems. It persists in roles, policies, tokens, trust relationships, pipelines, and interfaces. Responsibility lives in people, teams, ownership models, architecture diagrams, and memory. One survives reorganization cleanly; however, the other often doesn't. That mismatch is one of the central reasons organizations drift into an unsafe decision space, even when the people inside them are acting rationally.

This chapter is about that drift: how authority outpaces responsibility, why risk concentrates unevenly across the organization, and why the durable correction is not more reminders or better blame. It is digitizing responsibility: translating human expectations into enforceable system properties that survive in the same way authority does.

Authority Is Precise; Responsibility Isn't

Authority is a machine language. It is explicit, durable, and enforceable. Permissions are encoded precisely. A system can evaluate an instruction such as "Allow this identity to perform this action on this resource" with perfect clarity. That authority persists as long as the configuration exists.

Responsibility is a human language. It is expressed through ownership models, organizational charts, and informal expectations. Statements like "This team is responsible for that data" rely on shared understanding rather than structural enforcement.

Responsibility doesn't persist the way authority does. Authority lives in code and configuration. Responsibility evaporates when teams reorganize, people rotate, or projects change hands. When new actors or automation interact with a system, the meaning of "responsible" becomes ambiguous.

Local vs. Global Optimization

Modern systems are built through local decisions.

The person building or operating a service inhabits a bounded world. Their responsibility is to make that service function: to meet deadlines, deliver features, and keep systems running. From this vantage point, expanding authority is often the fastest path to success. If granting broader permissions allows the service to work and the deadline to be met, the local system has been optimized.

Constraint responsibility operates in a different world altogether.

Its concern isn't whether a single service functions, but how that service interacts with everything else. It asks how authority granted in one place creates new paths elsewhere, how small convenience decisions combine into systemic exposure, and how local optimizations reshape global risk.

As systems scale, these goals diverge.

You cannot reasonably ask builders to act as global optimizers without slowing them down. It's why organizations have historically relied on security reviews, architecture boards, and approval committees, which are manual attempts to inject global reasoning into local workflows. These mechanisms work only by trading speed for oversight. The tension is structural.

Rational Behavior as a Risk Factor

The divergence between authority and responsibility is often misread as negligence or poor judgment. It is neither, however.

When developers are measured by their velocity, reliability, and delivery, it is rational to expand their authority to remove friction. When teams are rewarded for unblocking use cases quickly, it is rational to bypass constraints that slow down progress. These decisions aren't reckless; they are aligned with the incentives in place.

The problem isn't that people choose speed over safety. The problem is that being fast and being safe are treated as competing goals.

In environments that govern possibility, this trade-off disappears. The system's structure changes so that moving fast is only possible within safe bounds. Developers don't need to choose between either velocity or security, because unsafe action variants aren't available options. This isn't a matter of developers needing more training or awareness. It is a matter of structural alignment. A Formula One track is a good analogy for when constraints are designed upstream, where the fastest way is also the safest way at those fast speeds.

Digitizing Responsibility

As systems become more automated, responsibility can no longer remain only social. In manual systems, organizations relied on human friction. Someone paused. Someone double-checked. Someone asked whether the access was still needed, whether the vendor should still be connected, whether the exception should still exist, or whether the deployment path was safe enough. That friction was imperfect, but it served a purpose. It created moments where responsibility could re-enter the system through human judgment. Those moments are disappearing.

CI/CD pipelines act continuously. Cloud permissions change through automation. SaaS integrations expand quickly. Agents call tools, update state, and inherit authority from the systems around them. Work increasingly happens at speeds where human review can't remain the main control. This creates a crisis of trust.

Organizations want automation because it gives them speed and scale. But they hesitate because they can't see the brakes. They can see the authority automation holds. They can't always see the responsibility that governs it. Digitizing responsibility is the answer to that problem. To digitize responsibility is to convert a human expectation into an enforceable system property. It means taking statements that used to live in requirements, review notes, architecture diagrams, policies, or team memory and encoding them into the systems that create and use authority.

Historically, many security responsibilities have been expressed as non-functional requirements, or NFRs: a system must protect sensitive data; temporary access must be removed after the launch; only approved workloads should communicate with the database; vendors must only access the data they need; pipelines must deploy only verified code; agents must not act beyond their assigned task. These requirements capture responsibility. But too often, they remain outside the system itself. They are reviewed in architecture discussions, checked during audits, interpreted by engineers, or rediscovered after incidents. The responsibility exists, but it remains social and procedural. That is the problem.

Authority already has a digital form. It lives in roles, policies, tokens, service accounts, trust relationships, deployment systems, and APIs. Responsibility often does not. It lives in documents, meetings, diagrams, approvals, and memory. As long as authority is digital and responsibility is social, the two will drift apart.

Digitizing responsibility means closing that gap. It turns NFRs from statements of intent into enforceable invariants. The requirement no longer says only what should be true. It becomes part of the control plane, workflow, token, identity, interface, or cryptographic condition that determines what the system can actually do. In this shift, responsibility gains the same properties as authority. It becomes explicit and enforceable. It persists through reorganization. It can be inspected and tested. It can travel with the authority it governs. The following table highlights this shift.

Table 5.1: NFRs translated into digitized responsibilities

NFR / Social Responsibility	Digitized Responsibility	Unsafe Trajectory Removed
"Only the application tier should talk to the database."	**Identity-based micro segmentation:** the database accepts connections only from verified workload identities through mutual TLS, not from arbitrary network locations.	Removes the lateral-movement path that allows a compromised peer to gain database access simply because it is nearby on the network.
"Temporary launch access should be removed after the event."	**Just-in-time ephemeral access:** access is minted through a brokered workflow with mandatory expiration tied to the change window.	Prevents the standing-authority path where temporary access quietly becomes a permanent risk.
"Pipelines should only deploy verified code."	**Attestation-based deployment:** deployment authority is granted only for artifacts with signed provenance and a valid workflow state.	Removes the deployment-bypass path that allows a pipeline to push unverified, substituted, or malicious artifacts.
"Vendors should only see the data they need."	**Scoped API brokering:** vendor integrations use dedicated identities and short-lived tokens restricted to specific APIs, datasets, and operations.	Prevents trust inflation where one vendor integration becomes a broad bridge into the internal environment.

NFR / Social Responsibility	Digitized Responsibility	Unsafe Trajectory Removed
"Service accounts must have an owner."	**Required identity metadata and lifecycle controls:** identity creation fails unless the owner, purpose, scope, and expiration or review cadence are defined.	Removes the orphaned-authority path where identities persist without lifecycle, accountability, or review.
"Agents should not overreach their task."	**Brokered tool use:** agents receive task-scoped sub-tokens only after a broker validates task context, tool scope, and expiration.	Prunes the agent-to-admin path where authority persists or compounds across unrelated task contexts.
"Sensitive data is for approved purposes only."	**Purpose-bound decryption:** keys are released only to workloads presenting a broker-minted purpose claim tied to approved use.	Breaks the possession-equals-access path where having a credential is enough to reach sensitive data.

The specific mechanisms will vary by organization. The important move is the conversion. A policy becomes a control-plane rule. A responsibility becomes a workflow state. An ownership expectation becomes required metadata. A temporary decision becomes an expiring authority. A trust assumption becomes a scoped interface. A human reminder becomes an enforced invariant. That is what it means for responsibility to become digital.

Consider the database example in the table earlier. In the social-responsibility model, the organization has an NFR that says only the application tier should talk to the database. A team may implement that expectation through firewall rules, subnet boundaries, or review processes. Those controls may help, but the responsibility still depends heavily on people remembering how the network is supposed to work. If someone moves a workload, opens a port, changes a subnet, or reuses access in a new context, the unsafe path may quietly reappear. The system then has to rediscover the violation later. It scans. It alerts. It opens a ticket. It asks someone to restore the intended boundary.

In the digitized-responsibility model, the expectation is encoded differently. The database accepts connections only from approved workload identities. Network location is no longer enough. A workload must present the right cryptographic identity, through the right path, under the right policy. The responsibility no longer depends only on remembering the intended topology. It is embedded into the communication grammar of the system. It is removed from the normal path.

This is the same pattern across the table. The point is not that every responsibility can be perfectly encoded. Nor is the point that documents, reviews, and human judgment disappear. Humans still decide what should be allowed, which suppliers should be trusted, what scope is appropriate, how long access should last, and which workflows should exist. But once those decisions are made, the system should not rely only on memory to preserve them. The decision should be encoded into the path of work. That is how responsibility survives at machine speed.

Digitizing responsibility also changes the role of security. Security is no longer only the team that reviews whether responsibility was honored after the fact. It becomes the function that helps translate responsibility into the systems where authority is created, widened, reused, and combined. This requires security to work closer to platform engineering, identity, cloud, product, data, and architecture teams. The question is not only whether a requirement exists. The question is where that requirement lives. Does it live in someone's memory, or in the control plane?

When Authority Outpaces Responsibility

The divergence between authority and responsibility doesn't appear all at once. It emerges gradually, through a sequence of decisions that are each reasonable in isolation. A permission is expanded to unblock a use case. An exception is granted to meet a deadline. An automation is given a broader scope to reduce operational toil. None of these actions is reckless. Each is taken in good faith, often under real pressure. What changes isn't the intent behind the decisions, but their cumulative effect.

Authority scales mechanically. Each new permission, integration, or automated action increases the system's decision space. Responsibility, by contrast, scales socially. It depends on shared context, institutional memory, and ongoing attention. As systems evolve, those social mechanisms lag behind. Over time, authority begins to outpace responsibility. This isn't a single failure point. It is a trajectory.

Early in a system's life, responsibility and authority may be closely aligned. The people who grant permissions understand their implications. The system's scope is limited, and interactions are few. As usage grows, that alignment weakens. New consumers appear, and new workflows depend on existing permissions. Automation begins to act on behalf of humans. Each step widens the gap.

Temporary exceptions persist longer than intended. Narrow permissions are reused in broader contexts. Constraints that once felt sufficient no longer match how the system is actually used. Responsibility remains nominally assigned, but its meaning becomes diluted as the system's reach expands.

From the control plane's perspective, nothing is wrong. All actions are valid, and all permissions are authorized. The system continues to operate exactly as configured. There is no obvious violation to point to and no single decision to reverse.

From the organization's perspective, however, the risk profile has changed. The system can now do more than it could before. It can reach further, affect more assets, and interact with more components than originally anticipated. This expanded decision space is rarely visible in one place. It is distributed across configurations, integrations, and automation logic that evolved over time.

This is why post-incident analysis so often reveals a familiar pattern: "None of these changes looked dangerous on their own." That observation is accurate and misleading. The danger wasn't in any individual decision. It was in the direction those decisions collectively pushed the system. Authority accumulated. Constraints eroded. Responsibility remained diffuse. The system drifted toward states that were never explicitly intended but were implicitly enabled.

A useful way to find this drift is to review high-authority roles, workflows, integrations, and automation paths and ask a few simple questions. Who created this authority? What problem was it meant to solve? Who depends on it now? What systems can it reach today? What responsibility is encoded into the system itself, rather than remembered by a team? If the answers are unclear, authority has likely outlived responsibility.

Once authority has outpaced responsibility, restoring balance is difficult. Constraints added later feel arbitrary. Capabilities that teams have come to rely on are suddenly called into question. Attempts to rein in such capabilities meet resistance, not because people oppose safety, but because the system has already adapted to the expanded possibility space. This is why governance that relies on periodic correction struggles to keep up. It is always acting against momentum.

Understanding this dynamic is critical because it reveals an uncomfortable truth: Risk isn't created by a single bad decision. It is created by a sequence of good decisions made without a durable mechanism to preserve responsibility as authority expands.

The Organizational Distribution of Risk

As authority outpaces responsibility, risk doesn't spread evenly across an organization. It concentrates. This concentration isn't random, and it isn't the result of individual negligence. It is the natural outcome of how authority, responsibility, and incentives interact as systems scale.

If you want to find where your organization is most exposed, you don't start with a vulnerability scan. You ask three questions, each pointing to a different mode of concentration that together form a diagnostic map of where structural constraints are most urgently needed.

Where has authority accumulated fastest?

Every organization has systems that started small and became load-bearing infrastructure through success. A platform team builds a shared deployment service. It works well, so other teams adopt it. Each adoption requires a small expansion, such as access to another account, data source, or environment. The service then becomes an integration hub. It is extended repeatedly because it is reliable, it unblocks others, and reuse is faster than building from scratch. This is rational behavior producing irrational outcomes.

Consider what happens with a legacy internal API, say one that handles customer identity. It was built years ago for a single product. Over time, it was integrated into onboarding flows, billing systems, support tools, analytics pipelines, and partner portals. Today, it touches hundreds of downstream applications. It has become the system's de facto identity backbone, a role nobody designed it for, nobody explicitly approved, and nobody can easily change.

The API is overconnected, overtrusted, and underconstrained. Its blast radius has grown with every integration, but responsibility for constraining that blast radius has never kept pace. Ownership of the API exists on paper. However, the governance of its cross-organizational impact doesn't.

It is how high-authority surfaces become risk magnets. They attract integration because they work. Each integration expands decision space. Responsibility diffuses as more teams depend on the system. The system becomes critical infrastructure without acquiring commensurate governance.

A practical diagnostic: Look at your environment and ask which systems would cause cascading failure if compromised, not because of what they store, but because of what they connect. Those are your risk magnets. If you cannot name the person who governs their cross-boundary blast radius (not just their uptime), you have found a gap between authority and responsibility.

Where is change fastest and pressure highest?

Risk also concentrates where delivery pressure is greatest, because pressure is where responsibility fails first. Teams under deadline are granted broader authority to move quickly. Constraints are relaxed "temporarily." Automation is given a wide scope to reduce operational toil. These isolated decisions make sense in the moment. What lags is the durability of the reasoning behind them, which plays out in a pattern every practitioner recognizes.

A team is given elevated access to a core database for a weekend launch. The launch succeeds, and the team pivots to the next priority. The access remains, not because anyone decided it should, but because no one was assigned to revoke it, and the system has no mechanism to expire it automatically. Six months later, the environment carries the accumulated authority of every urgent moment that came before, long after the urgency has passed.

The same pattern appears with automation. A platform team creates a single deployment account with broad permissions because it is easier to maintain than 50 scoped accounts. The decision is pragmatic, and it creates a single identity whose authority spans the entire production environment, governed by whoever last edited the Terraform file.

And increasingly, this pattern extends to AI. An autonomous agent needs access to a data lake. Rather than creating a purpose-bound identity and going through a security review, which may take days, someone attaches the agent to an existing legacy role that is already trusted. The agent then inherits authority that was granted under entirely different assumptions, to a different actor, for a different purpose.

The common thread isn't negligence. It is that responsibility fails under pressure, while authority persists after the pressure passes. A builder will always choose the general or inherited path if the precise path takes two extra days. The system's design, not the builder's character, determines whether that choice creates lasting risk.

A practical diagnostic: Look at where your teams have the shortest deadlines and the broadest access. Look at where "temporary" is a word that appears in access justifications but not in access expiration policies. That is where authority has outrun responsibility, and where structural constraints would have the highest immediate impact.

Where have trust boundaries quietly expanded?

The first two questions focus on where authority gathers and where it grows under pressure. This third question focuses on something subtler: How trust compounds across boundaries until the boundaries themselves become meaningless.

In modern environments, systems are rarely isolated. They are connected by cross-account roles, API integrations, service-mesh relationships, and third-party access grants. Each connection is a trust relationship, and trust relationships are among the most durable forms of authority in any system.

The problem is that trust boundaries aren't static, but the authority granted under them often is. Consider a cross-account trust relationship: Account A (development) is allowed to assume a role in Account B (production) to move logs. The relationship is narrow and well-understood when it is created. Over time, Account A's environment changes. New services are deployed, new automation runs, and new developers join. The trust relationship persists unchanged, but the set of actors who can exercise that trust has expanded far beyond the original intent. What was once a narrow audited path between two known systems has become an implicit guarantee that whatever runs in Account A can reach into Account B.

Consider third-party integrations. You grant a code-quality scanning platform an API key to your private repositories. The integration is reviewed and approved, but the vendor's security posture is now your security posture. Their breach becomes your exposure. Their employee with access to customer data now has a path, however indirect, to yours. The trust boundary between your organization and theirs has been bridged by a single credential that nobody reevaluates after the initial review.

Even within a zero-trust architecture, this pattern persists. When a trusted identity is granted access to an entire API rather than a specific set of operations, you have moved the "flat network" problem up the stack into the identity layer. The network perimeter may be gone, but the identity perimeter has inflated to fill the same space. This phenomenon is what I call *trust inflation.*

Trust Inflation

Just as monetary inflation devalues currency by increasing its supply without increasing its underlying value, trust inflation devalues the security of an organization by expanding the supply of trust, through inherited roles, persistent access, and cross-boundary relationships, without a corresponding increase in structural constraint.

Trust inflation is the organizational mechanism that produces the kind of bridging identity described in Chapter 2, where the IAM role grew through legitimate approved expansions until it connected what the organization was designed to keep apart. That role wasn't an anomaly. It was the predictable output of an environment where trust is cheap to grant and expensive to constrain. Trust inflation isn't just an expansion of access. It is an expansion of the set of trajectories that become reachable through a single identity.

Trust inflation is also cumulative. Each inflated trust relationship makes the next one easier to justify. "This account already trusts that one, so adding one more integration won't matter." "This role already has broad access, so attaching one more service won't change the risk profile." The marginal cost of each expansion feels negligible. The aggregate cost is an organization where the perimeter, whether network or identity, no longer offers meaningful resistance to lateral movement.

Note: The phrase "inflation of trust" has been used in a different context by König and Sumpf (2021), who examine how digital platforms overextend user trust as a mechanism for managing uncertainty. The concept of trust inflation in this book addresses a different phenomenon (the accumulation of organizational trust relationships, cross-account roles, vendor integrations, and inherited identities), which expand the supply of trust without corresponding structural constraints, thereby devaluing the isolation boundaries the organization designed.

A practical diagnostic: Map the trust relationships that cross your most important boundaries, like account boundaries, environment boundaries, organizational boundaries, and vendor boundaries. For each one, ask, "Does the set of actors who can exercise this trust today match the set that was contemplated when the trust was established?" If not (it almost never does), you have found trust inflation. The gap between the original intent and the current reality is a measure of how much your isolation boundaries have quietly eroded.

The three questions you've asked don't produce a vulnerability list. They produce something more useful. They show you where structural constraints are most overdue:

- **Risk magnets** tell you where a blast radius has grown beyond any one team's natural ability to govern it.
- **Pressure points** tell you where builders are repeatedly being forced to choose between speed and safety, because the system offers no path that reliably delivers both.
- **Inflated trust boundaries** tell you where isolation has quietly eroded and where identity architecture needs structural tightening.

These concentrations often overlap. Risk magnets, pressure points, and inflated trust boundaries are not separate failure modes. They are the same structural fact seen from different angles: authority is easy for systems to preserve and expand, while responsibility is easy for organizations to dilute. A role remembers what it can do. A token remembers where it is accepted. A trust edge remembers whom it connects. The system does not remember the launch pressure, the temporary rationale, the departing owner, or the boundary the organization meant to preserve unless that responsibility has been encoded into the path itself.

The durable correction isn't more reminders, more attestations, or better blame after the fact. Those may create visibility, but they do not make responsibility survive. Responsibility must be digitized: translated from expectations, diagrams, policies, and team memory into enforceable properties of workflows, identities, interfaces, tokens, and control planes. Temporary authority must collapse when temporary work ends. Cross-boundary trust must remain scoped to the reason it exists. High-authority workflows must carry their purpose, owner, limits, and terminal state with them.

Modern security cannot govern dynamic systems by relying on people to remember what authority was meant to mean. It must design environments where responsibility travels with authority, constrains it as it moves, and survives the organizational changes that would otherwise erase it. Authority already has a durable system form. Responsibility must become just as durable as the authority it governs.

CHAPTER 6

Redefining How Security Operates Inside Organizations

Security doesn't usually fail inside organizations because the policy is unclear. It fails because, under pressure, the supported path is slower than the workaround:

- A team knows the approved deployment flow, but the custom pipeline is faster.
- An engineer knows the access request should go through the right channel, but the shared role already works.
- A platform team knows the safer pattern exists, but it is incomplete, poorly documented, or missing the one feature the business actually needs.

In practice, security doesn't succeed or fail at the level of policy. It succeeds or fails at the level of how work gets done: which paths are easy, which are painful, which are supported, which are trusted, and which quietly fall into disuse. The daily operating environment, not the rulebook, determines whether security ideas spread or stall. That is why redefining security as constraint design isn't enough:

- A constraint that nobody uses isn't yet a control.
- A paved road that teams bypass isn't yet a path.
- A secure default that gets overridden under deadline isn't yet the default in any meaningful sense.

Security cannot remain primarily a gatekeeper that reviews decisions after they are made. It has to become a designer of systems, defaults, workflows, and incentives that shape decisions before they arise. Once security takes on that role, incentives stop being an external force to fight and become a design input. Defaults, friction, and ease of use become security primitives, because they determine which paths are chosen under pressure and which are ignored.

Once security operates that way, it has to be measured differently:

- Not by how many policies exist
- Not by how many exceptions are reviewed
- Not even by how many controls are deployed

Security must be measured by propagation: by whether secure designs are reused, extended, trusted, and chosen as the normal way work gets done.

This chapter examines this shift in three dimensions: the role security plays, the levers it uses, and the measure by which it knows whether the model is actually taking hold.

From Gatekeeper to System Designer

If security is no longer primarily about enforcing rules after intent has formed, then its role must change. In many organizations, security still operates as a gatekeeper. Teams build something, request approval, wait for review, and receive either a yes, no, or list of required changes. This model assumes that security creates value at decision points: reviews, sign-offs, exceptions, and approvals. That assumption no longer holds.

In most organizations, security decisions are made long before a request ever reaches a security queue. They are made when a platform is designed, when a default is chosen, when an interface is exposed, when a workflow is standardized, and when authority is scoped for reuse. By the time a team asks,

"Is this allowed?", the environment has already shaped what is possible, which is why the shift here is not simply "do reviews earlier." It isn't just a call to move design reviews, threat models, or sign-offs further left in the lifecycle. Those practices operate on individual changes. The deeper shift is that security must help design the environment in which those changes are made. That is the move from *gatekeeper* to *system designer*.

As a system designer, security focuses less on evaluating requests and more on shaping the supported paths that teams use every day. The goal isn't to review every action. The goal is to make most actions safe enough that they don't require review at all.

Consider a deployment pathway.

In a gatekeeper model, each team assembles its own deployment workflow, chooses its own identities, wires its own secrets, and then asks security to review the resulting design. Security evaluates after the path already exists.

In a system-designer model, security helps ship one supported deployment path: validated artifacts, bounded deployment identities, reliable auditability, environment separation, rollback behavior, and clear progression rules already built in. Teams don't begin by inventing their own path. They begin by choosing from a supported one, which changes the day-to-day experience:

- Instead of asking "Is this deployment allowed?", teams ask, "Which supported deployment path should I use?"
- Instead of requesting exceptions, they choose among predefined options that already encode security requirements.
- Instead of waiting for approval, they move forward confidently because the system already defines what acceptable execution looks like.

This doesn't mean security becomes invisible because it matters less. It becomes less visible because its influence is expressed earlier and more durably.

The practical shift is straightforward: Security stops spending most of its time catching unsafe decisions, and starts spending more of its time designing safe decisions into platforms, templates, identity patterns, and workflows.

- If most security work appears as tickets, reviews, and exception requests, security is still operating primarily as a gatekeeper.

- If most security work appears as changes to platforms, paved workflows, templates, and identity patterns that many teams then reuse, security is operating as a system designer.

This also changes what the security function needs. It needs people who can design and ship platforms, not only review them. It needs durable interfaces with platform, developer enablement, and infrastructure teams. It needs to think in terms of adoption, product quality, and reusable building blocks, not only control coverage. Most importantly, this shift aligns security with how modern organizations actually work.

When security designs the systems people use every day, it stops competing with delivery and becomes part of delivery. It is what effective security looks like inside a mature organization.

Incentives as a Security Primitive: A Choice Architecture Approach

Once security becomes a system designer, a subtler question comes into focus: *How do secure designs actually win inside everyday work?*
Security doesn't operate in a vacuum. It operates inside environments shaped by deadlines, delivery pressure, operational toil, and competing goals. In that environment, the question teams ask is rarely "What is allowed?" The real question usually is: *What is easiest?*

It is why incentives matter so much. Authority scales mechanically. Responsibility scales socially. But incentives determine which of the available trajectories are actually chosen under pressure.

A useful way to frame this is through *choice architecture.* This term was popularized by Richard Thaler and Cass Sunstein in their book, *Nudge: The Final Edition.* The basic insight is simple: There is no neutral environment. Systems always make some paths easier than others. In security, that means the real design question isn't whether people care about doing the right thing. It is whether the environment makes the safe path the natural path.
Three levers determine how incentives operate inside a system:

- *Defaults* shape what becomes normal.
- *Friction* shapes what feels exceptional.

- *Inline governance* shapes what is expressible at all.

Defaults as Organizational Gravity (Redefining What Defaults Mean)

Secure defaults have long been a rallying cry in security circles. In most contexts, it means something narrow and technical:

- Encryption enabled by default
- Public access disabled by default
- Least-privilege permissions as the starting point

While these are important practices, they aren't the redefinition I am arguing for. Defaults aren't just settings. They are organizational gravity. They determine which systems get adopted, extended, and trusted under pressure. They shape what happens when nobody stops to think. Constraint design defines what is possible, while defaults determine what becomes normal.

When the default deployment path embeds logging, identity scoping, network boundaries, and auditability, teams rarely "choose" security. They choose the fastest way to get work done, and security comes with it. Over time, that path becomes infrastructure. It is reused, copied, and assumed.

But when the default path is incomplete or permissive, teams fill in the gaps themselves. They create custom roles, ad-hoc pipelines, and bespoke access patterns. These don't arise from malice or ignorance. They arise because the insecure path was easier to extend than the secure one. This is the mechanism through which trust inflation takes hold. Every custom integration, every bespoke access pattern, every ad-hoc pipeline is a trust relationship that the organization didn't plan and will struggle to constrain.

The critical design question isn't "Is this configuration secure by default?" It is "If nobody pauses to reconsider anything, what architecture will this organization naturally converge toward?"

If the answer is "a constrained, observable system," the defaults are working as a strategic security control. If the answer is "a permissive system that relies on later correction," security is relying on vigilance rather than design. That is the redefinition.

Secure defaults aren't about preventing mistakes. They are about determining which security architecture spreads by inertia. When defaults are treated this way, at the level of platforms, identity models, and supported workflows, they stop being technical settings and become the primary mechanism through which constraint design propagates across the organization. It is why defaults aren't a detail. They are a strategic lever.

Friction as a Directional Signal (Redefining What Friction Is For)

Security often treats friction as a defect. If something slows a team down, it is seen as a failure of enablement. Under delivery pressure, this instinct becomes stronger: approval steps are removed, checks are removed, and anything that might block velocity is removed.

Friction isn't inherently bad, however. It is one of the few levers that security controls. The traditional mistake is distributing friction evenly. When both safe and unsafe paths feel equally burdensome, teams will optimize for speed and convenience, not constraint.

The redefinition is more precise. Friction shouldn't be eliminated. It should be placed deliberately at the edges of decision space expansion.

In well-designed systems, the safe and supported path is fast, reliable, and requires low effort. The path that expands authority, introduces novelty, or increases the blast radius requires intentional effort. This isn't punishment. It is a signal.

Consider the evolution of GitHub workflows. In most organizations today, merging code through a pull request with automated checks is the path of least resistance. Direct pushes to production branches are possible but require elevated privileges and deliberate override. The friction isn't random. It is concentrated at the point where risk increases.

In financial systems, pre-trade risk engines operate similarly. Executing trades within established limits is seamless. Exceeding position or credit thresholds require manual intervention or explicit override. Friction appears exactly where exposure increases. These systems don't attempt to eliminate friction. They use it to shape behavior, and that is the shift.

Security must stop trying to remove friction everywhere and instead use friction as a structural cue. The safest patterns should feel effortless. Expanding decision space should feel deliberate. Novelty should be visible.

When friction is aligned with constraint design, it doesn't feel like security interference. It feels like how the system works. Teams move quickly within governed paths because those paths are engineered to be smooth. They slow down only when attempting to widen the system's authority surface. That isn't bureaucracy. It is architectural intent expressed through experience.

When friction is placed intentionally, it reinforces structural constraints instead of competing with them. It becomes a directional force that nudges systems toward stability without constant oversight.

In environments moving toward autonomy, this shift matters even more. Machine-driven systems will relentlessly exploit the lowest-resistance path available. If the lowest-resistance path is also the safest, incentives and constraint design are aligned. That is friction used as a security primitive.

Defaults determine what becomes normal. Friction determines what feels deliberate. Together, they shape the paths teams choose under pressure. However, there is a third lever that operates at a deeper layer; not shaping which path is chosen, but defining which paths exist at all.

Inline Governance (Redefining Where Inline Feedback Operates)

Many organizations think they are shifting left because security shows up earlier. IDE warnings flag risky code. CI pipelines fail insecure configurations. Policy engines block deployments that violate rules. These mechanisms are useful. They reduce the delay between mistake and correction. This is still feedback, where the unsafe thing was expressible first. The system then reacted.

Inline governance operates at a different layer. Instead of reacting to unsafe expressions, it defines which expressions are valid in the first place. Unsafe states aren't flagged; they are unrepresentable.

Modern cloud-native systems already demonstrate this distinction. In Kubernetes, admission controllers operate inside the API server which is the control plane itself. When configured properly, they don't warn that a container is privileged or running as root. They reject the configuration before it ever becomes cluster state. A pod without resource limits, a workload from an unapproved registry, or a container requesting forbidden capabilities is simply not a valid object in the system. The cluster never enters that state.

Similarly, in Amazon Web Services (AWS), Organizations, service control policies (SCPs) don't generate alerts after risky API calls occur. They redefine what an account is capable of doing. If a policy denies the creation of public S3 buckets or restricts IAM user creation, those actions aren't temporarily flagged. Instead, they are structurally disallowed at the control-plane level. Even administrators cannot bypass them through standard means. The grammar of authority has been narrowed.

The same pattern appears in GitHub with protected branches. Direct pushes to a protected branch aren't "discouraged." They aren't permitted unless strict conditions are met. Required reviews and status checks are enforced as structural preconditions for state transition. The repository cannot enter an unsafe state by accident.

These examples share a common property: Governance is embedded where intent is translated into state. This is the redefinition.

Inline governance doesn't wait to see what happens. It shapes what can happen. It operates inside the control plane, the layer where identities, permissions, and state transitions are defined. It survives automation because it doesn't depend on reflection. It constrains the decision space before it can be explored. This distinction becomes decisive in autonomous systems.

An autonomous agent doesn't hesitate when confronted with a warning. It optimizes within the authority it has been given. If unsafe options exist, they will eventually be exercised. Inline feedback cannot keep pace with machine-speed exploration of possibility space. Inline governance can, because it removes unsafe possibilities entirely.

Security becomes ready for autonomy not when alerts arrive faster, but when the system's control plane defines intent so tightly that unsafe trajectories cannot form at all.

Designing for Propagation, Not Compliance

If security is constraint design, and incentives shape which paths are chosen, then the focus naturally turns to evaluation: whether our security designs are producing the outcomes they were built to achieve. Traditionally, security has measured success through compliance and coverage:

- How many policies are written
- How many systems are scanned
- How many exceptions are approved
- How quickly incidents are detected and resolved

These metrics describe activity. They don't describe adoption.

A control that exists but is rarely used isn't a strong control. A pattern that requires constant enforcement isn't stable. A rule that generates frequent exceptions isn't aligned with the system it governs.

In our constraint designer model of security, the primary measure of success is different. It is propagation. *Propagation* is the design of security as a *self-distributing trait.* It is the mechanism by which a secure decision made automatically travels to every corner of the environment. If Choice Architecture (refer back to 6.2 Incentives as a Security Primitive: A Choice Architecture Approach) ensures that the right path is chosen at the start, propagation ensures that safety is inherited by everything that follows.

Propagation asks:

- Are secure patterns reused without being mandated?
- Do new teams default to the supported path?
- Are unsafe patterns declining naturally and not just through an audit?
- Does expanding the organization reinforce constraints or erode them?

A security design that propagates doesn't need constant supervision. It becomes part of the organization's infrastructure. This reframes how security should measure itself.

Security no longer asks How many violations did we catch?", it begins asking, "How many decisions did we eliminate by design?" Instead of asking "How many exceptions were requested?", it asks "Why was an exception necessary at all, and how do we redesign the system so it isn't?" And instead of asking "Did teams follow the rule?", it asks "Did the teams choose the supported path without being told?"

Those are harder questions that require looking at patterns over time, not single events.

In environments approaching autonomy, this shift becomes even more critical.

When systems operate at machine speed, reactive metrics lose meaning. Mean-time-to-detect and mean-time-to-respond describe how quickly you react after a trajectory has begun. They say little about whether unsafe trajectories were structurally possible.

In an autonomous world, stronger indicators include:

- The percentage of workflows that rely exclusively on supported, trajectory-governed platforms

- The ratio of scoped roles to custom high-privilege identities
- The frequency with which structural limits are refined rather than bypassed
- The rate at which insecure legacy patterns disappear without enforcement campaigns
- The degree to which new services inherit secure defaults automatically

These metrics are quieter. They measure architecture, not alarms. They ask whether security designs are spreading faster than insecure ones. When read together, they are also a measure of trust inflation. Each metric tracks a different dimension of whether unconstrained trust is accumulating faster than structural constraints can govern it. Workflows move off the supported path. Roles expand beyond scoped patterns. Bypasses become routine. Legacy patterns persist. New services fail to inherit defaults. These aren't different problems. They are different expressions of the same organizational dynamic: trust expanding without corresponding constraint.

Propagation and trust inflation are the same phenomenon measured in opposite directions. When constraints propagate faster than trust inflates, the organization is gaining ground. When trust inflates faster than constraints propagate, it is losing ground, regardless of what the compliance dashboard says.

This is the real test of security inside an organization. A control has not succeeded because it exists. A policy has not succeeded because it was approved. A paved road has not succeeded because it was announced. Security succeeds when the constrained path becomes the path people actually choose under pressure, and when each new team, service, workflow, and system inherits safer defaults than the one before it.

That is why propagation, not compliance, is the signal that the operating model is ready for what comes next. Before autonomy can be governed safely, constrained patterns must already be spreading faster than unconstrained trust. Governance cannot remain optional, advisory, or downstream. It must become structural before machine-speed execution arrives.

CHAPTER 7

Governance Before Autonomy

Autonomy does not enter a neutral environment.

It enters the decision space the organization has already built: the roles already granted, the workflows already normalized, the interfaces already exposed, the tools already connected, and the trust relationships already allowed to spread. An autonomous system does not pause to ask whether those paths reflect the organization's current intent. It uses what is reachable.

That is what makes autonomy unforgiving. It doesn't merely add a new actor to the enterprise. It turns existing authority design into executable behavior. If constrained paths have propagated, autonomy can scale them. If broad roles, inherited trust, standing permissions, and off-road workflows have propagated, autonomy can scale those as well. The system will not treat them as historical accidents or temporary exceptions. It will treat them as available operating paths.

A human operator may hesitate before using broad authority in an unusual way. A team may pause before crossing a sensitive boundary. An experienced engineer may sense that a technically valid action is unsafe in context. That informal caution has carried more weight than most security models admit. It has acted as a hidden compensator for weak structure: the pause, the doubt, the conversation, the escalation, the uneasy feeling that a permitted move is still the wrong move.

Autonomous systems do not supply that compensating friction. They explore the decision space made available to them, compose actions through the interfaces they can reach, and execute within the authority they have been given. The result is a shift in the unit of security: from *action* to *trajectory*. The question is no longer only whether an individual action is authorized. The question is whether the actor should have been able to reach that action, in that sequence, with that degree of freedom, at all.

This is why autonomy invalidates downstream governance as the primary model. Detection, monitoring, and response still matter, but they operate after the system has begun to move. In an autonomous environment, that intervention may be too late. A system can execute thousands of authorized but unintended actions before a human can understand the sequence, let alone intervene.

The issue is not that every autonomous system is dangerous. The issue is that autonomy removes the pause that once made weak governance survivable. If decision space remains broad, if authority can still be inherited casually, if high-impact workflows can still bypass the supported path, and if trust can still expand without collapse, then autonomy turns those design debts into executable behavior.

This chapter is about the threshold where governance can no longer remain advisory, after-the-fact, or dependent on human hesitation. Once systems act at machine speed with inherited authority, governance must move into the design of the environment itself. It must define which behaviors must be impossible before execution begins, where structural intent must be encoded, and why the control plane is the only place that can hold those constraints at speed.

When Governance Must Become Structural

Not every system requires the same level of structural governance at the same urgency. Some systems can still be governed through review, monitoring, guardrails, and response. Others cross a threshold where those controls remain useful, but can no longer be the primary safety model. That threshold is reached when the following five conditions begin to hold:

- **Speed Threshold** is the point where execution moves faster than governance can intervene. A person may still review the design, investigate the alert, or approve a policy later, but they are no longer on the critical path of the decision. When the system can act before oversight can matter, the constraint has to exist at the point of execution.
- **Scale Threshold** appears when the volume of transitions exceeds human cognitive bandwidth. A handful of high-impact changes can be reviewed manually; thousands of identity changes, API calls, deployments, SaaS integrations, and agent actions can't. At that scale, policies become advisory unless they are encoded into defaults, workflows, interfaces, and control planes.
- **Irreversibility** is the point where correction stops being a meaningful security model. Data can't be un-leaked, a harmful external message can't always be unsent, a physical action can't always be rolled back, and a minted credential may create effects that survive revocation. In those systems, detection and response still matter, but they can't be the primary way safety is achieved.
- **Context Fragility** appears when the system may be wrong about the situation it is acting in. An autonomous system can misread a user request, trust poisoned context, retrieve the wrong record, confuse a draft with an approval, or treat an inferred goal as if it were authorized.
- **Compositional Drift** is the point where risk lives in the chain rather than the individual permission. A system may read data, draft output, invoke a tool, update state, and take external action, with each step authorized on its own. If those valid steps can combine into an unsafe

outcome, the organization is no longer dealing with a simple access-control problem; it is dealing with a trajectory problem.

I have seen the same conditions mark the transition across domains as different as aviation, medical devices, and autonomous software systems. Let me summarize again in the following table:

Table 7.1: Conditions where traditional governance fails

Condition	What It Means	Traditional Governance Failure	Trajectory Governance Requirement
Speed Threshold	Execution moves faster than review, triage, or approval can matter.	Detection arrives after the system has already acted. The human becomes a historian, not a gate.	Constraints must exist at the point where the action is taken.
Scale Threshold	The volume of transitions exceeds human cognitive bandwidth.	Human oversight cannot keep pace with thousands of identity changes, API calls, deployments, integrations, and agent actions.	The safe path must be the normal supported path, not an exception that requires constant review.
Irreversibility	Some outcomes cannot be meaningfully undone once they occur.	Response becomes cleanup after the consequence: data cannot be un-leaked, external actions cannot always be unsent, and some credentials create effects that survive revocation.	Unsafe states must be unreachable before execution, not merely remediated afterward.

Condition	What It Means	Traditional Governance Failure	Trajectory Governance Requirement
Context Fragility	The system may misunderstand the situation it is acting in.	An autonomous system may misread a request, trust poisoned context, confuse a draft with an approval, or treat an inferred goal as authorized.	High-impact actions must depend on verified context, explicit workflow state, scoped authority, and approval conditions the model cannot invent for itself.
Compositional Drift	Risk appears in the sequence, not the individual action.	Each step may be authorized on its own, but the chain can still produce an unsafe outcome.	Constrain the workflow so low-risk actions cannot connect into high-impact paths without required state transitions.

Most modern cloud-native environments already meet at least two of these conditions. Many increasingly meet most of them.

Beyond it, behavior-centric security collapses into forensics. Detection, audit, and response are still important, but they no longer govern outcomes. They explain what happened, support recovery, and provide learning signals. They don't reliably prevent high-speed systems from traversing unsafe paths that were already reachable.

The operative question, therefore, changes. It is no longer: How do we detect bad behavior quickly enough? It becomes: What behavior must be impossible by design before execution begins?

From Behavioral Monitoring to Structural Intent

A useful way to see structural governance clearly is to look at a domain where it is not optional. Consider an automated drug-delivery system. An overdose can't be meaningfully rolled back after the fact. A delayed alert can't undo harm. It is why such systems don't rely primarily on monitoring and response to remain safe. They encode safe intent structurally.

A clinician may request a dosage. But the device doesn't accept arbitrary instructions simply because an authorized user issued them. Therapeutic ranges, rate limits, cumulative-dose ceilings, and interlocks are built directly into the control logic. Unsafe delivery isn't a valid expression that the system will later detect and correct. It is not a valid instruction. The system still automates. The human still exercises judgment. But the range of reachable trajectories is bound before execution begins.

Digital systems increasingly need the same design logic. The point isn't that software systems are identical to medical devices. The point is that once execution is fast enough, consequential enough, and difficult enough to supervise, safety no longer comes primarily from watching behavior more carefully. It comes from encoding intent into the system, so that certain dangerous instructions are never valid to begin with.

For any high-impact workflow, the test is simple:

Can the system express the unsafe instruction?

If the answer is yes, then you're still relying primarily on detection, correction, or human intervention. If the answer is no, then you have begun to achieve structural governance. That is the conceptual bridge into the control plane.

In digital systems, the place where instructions become state, authority, and reachable next moves isn't the runtime surface alone. It is the control plane that defines what identities can exist, what transitions are valid, what interfaces are exposed, and how workflows progress. That is where structural governance must live.

Governing Decision Space Through the Control Plane

Decision space cannot be reliably governed solely at runtime. Most modern security programs already talk about "guardrails," but the real question is where those guardrails live. In many organizations, they still live in the execution plane in runtime policy checks, CI/CD failures, scanners that block or revert changes, alerts that trigger rollback, and feedback systems that attempt to catch unsafe intent after it has already been formed. These mechanisms can be useful. They are not enough, though, under autonomy.

Appendix A frames the broader first-principles problem: Agents are decision systems, so the security question isn't only what they can access, but which trajectories their authority allows them to pursue.

At machine speed, execution-plane guardrails become reactive artifacts. By the time a rule fires, the system may already have composed the action with others, delegated it, repeated it, or propagated its effects across multiple boundaries. Feedback that depends on pause and reconsideration is a weak model when the actor doesn't pause. It's why the shift required here is not merely temporal.

True shifting left isn't only occurring earlier in time. It also occurs higher in the stack. Governance must move from the execution plane into the control plane. It means moving from catching unsafe actions after they are expressed, to defining which actions, identities, and sequences are expressible at all. This movement becomes especially visible in the way many organizations are currently authorizing agents.

The default pattern isn't a deliberate autonomy model. It is delegated identity. The agent inherits a user's token, session, or role, and is therefore able to do whatever that human principal can do. This action is so attractive because it's fast. It is also dangerous because it turns a narrow task into a broad blast radius:

- A scheduling agent doesn't need the full authority of an employee.
- A code-review agent doesn't need access to HR systems.
- A support agent doesn't need unrestricted external communication simply because the person who configured it did.

Having a delegated identity confuses *who the principal is* with *what the system should be allowed to decide*. It encodes standing power, not bounded intent.

When the agent is prompt-injected, misinterprets context, or composes an unsafe sequence, the blast radius becomes everything the inherited identity could do, not everything the workflow should've made reachable.

Teams often try to fix this by moving to narrower role-based access control (RBAC) models or richer relationship graphs. That can improve static resource scoping. But it still misses the unit that matters most under autonomy: the sequence.

Risk is rarely a single permitted call. It is a chain:

read sensitive data → draft output → invoke tool → take external action

If authorization is evaluated only as a static check at each step, the system can remain "authorized" while still producing an unsafe trajectory. This is why governance must move into the control plane as a *stateful pathway*, not merely a permission list.

In this model, authority is minted for a workflow, not inherited as ambient standing power. High-impact actions become reachable only from specific states and only after required prerequisites are satisfied. Authority collapses when the workflow reaches its terminal state.

A simple sequence makes the difference concrete:

Gather: Read the relevant records.

Draft: Produce the candidate output.

Approve: Require explicit human review for external or high-impact action.

Execute: Only allow the action if the system is in the correct state, within scope, and within time.

In that model, "send external email" isn't something you try to block late, after the agent has already decided to do it. It is something the system cannot reach until the workflow has progressed through the states that make that action valid.

The guardrail is no longer a late-stage bouncer. It's a control-plane interlock, where there's …

- no more runtime warnings,
- no more policy attached to already-broad identities, and
- not a more elaborate review process around unrestricted power.

It's where there are control-plane primitives that make authority purpose-bound, interfaces curated, transitions stateful, and collapse automatic. From this perspective, the governance requirements of autonomy become much clearer:

- Identities must be purpose-bound and not generalized.
- Interfaces must expose curated actions and not a raw control-plane power.

- Permissions must encode progression and not just possession.
- Trust relationships must be explicit, minimal, and terminal wherever possible.

A deeper technical breakdown of why delegated identity, static RBAC, and relationship models each fail in different ways for agentic systems can be found in Appendix D. The point here is simpler and more urgent: If governance remains primarily in the runtime layer, then autonomy will certainly outrun it. If governance moves into the control plane, autonomy can then operate inside a bounded authority, rather than in an open-ended authority surface.

Organizations that have already redefined security as constraint design, embedded that design into the control plane, and aligned incentives so that constrained paths propagate will find that autonomy amplifies what they already know how to govern.

Organizations that haven't redefined security will experience autonomy as volatility, fragility, and surprise. Autonomy isn't the disruption. *Autonomy is the reveal.*

It reveals whether safe patterns have already propagated. It reveals whether broad authority, inherited trust, and off-road workflows have already become normal. In addition, it reveals whether the organization has built real interlocks or only faster ways to notice that the system moved into dangerous territory.

PART III

Institutionalizing Structural Constraints and Trajectory Governance

Once autonomy can move at machine speed, governance can no longer depend primarily on detection, vigilance, or pause. If unsafe trajectories remain reachable, the system will eventually traverse them. Part III is the engineering response to this reality.

In the early days of industrial manufacturing, safety was scaled through warnings, training, and operator attentiveness. Mature factories learned that this wasn't enough. They built *interlocks,* which are mechanisms that made dangerous sequences impossible to execute in the first place. A machine couldn't energize with the guard open. A step couldn't occur out of order. Safety stopped depending mainly on memory and started depending on structure.

Digital systems now face the same shift. For years, security has been scaled through standards, reviews, approvals, scanners, and alerts. Those practices still matter, but they aren't sufficient in environments where control planes continuously reshape authority, workflows move under pressure, and automation or agents can act faster than humans can meaningfully intervene. In those systems, safety cannot remain a downstream activity. It must be built into the architecture of how the system works. Building the interlocks for modern software is what this part covers.

Structural constraints remove unsafe variants from the set of valid states. *Trajectory governance* goes further. It constrains which transitions and sequences are reachable at all. The goal isn't to watch the system more carefully after it moves. The goal is to design the system so certain dangerous futures never become reachable in the first place.

Chapter 8 discusses how to encode these interlocks into control-plane primitives like transition models, purpose-bound authority, curated actions, progression rules, and paved roads. Chapter 9 shows how to operate those interlocks under pressure by monitoring the boundary of decision space, detecting drift early, and turning incidents into constraint patches. Chapters 10 and 11 move from engineering to institution-building, focusing on how a chief information security officer (CISO) leads this shift, and how an organization makes it real without slowing the business down.

Part III covers the shift from describing risk to deleting reachable failure classes. It moves from reacting to unsafe behavior to engineering systems in which unsafe trajectories are structurally harder, narrower, or impossible to form.

CHAPTER 8

Encoding Structural Constraints

Pick one workflow. Delete one unsafe future. Then make that deletion durable.

That is how trajectory governance becomes engineering. It doesn't begin with an enterprise-wide map of every possible control, nor with a catalog of every risk the organization may someday want to constrain. It begins with one high-impact workflow that already matters: a path where authority is created, speed is necessary, pressure is real, and failure has a meaningful blast radius

The first move isn't to watch that workflow more carefully. The first move is to change what the workflow can express. A production deployment without a validated artifact shouldn't be merely discouraged, detected, or rolled back. It should be unreachable. Privileged debugging without an active incident shouldn't be a policy violation waiting to be caught. It should be impossible through the normal path. An agent executing an external action without approval state shouldn't depend on a warning, a log, or a human noticing in time. The system shouldn't be able to enter that trajectory.

The prior chapters argued that unsafe futures form through authority, sequence, interface, and trust. This chapter shows how to encode constraints at those points, thereby removing the unsafe future from the system's design.

This work has a repeatable pattern. First, define the workflow as a transition model, so the control plane can distinguish a valid next move from an unsafe jump. Then build the authority layer, so power is minted only for a purpose, in the right state, with the right scope, and with a terminal condition. Then expose curated actions, so teams and agents act through intent-level interfaces rather than raw control-plane power. Then enforce progression rules, so valid actions can't compose into invalid sequences. Finally, package the whole system as a paved road that teams can use under pressure because it is faster, clearer, and more reliable than going off-road.

Throughout this chapter, I use the production deployment path as the primary example. It connects directly to the CI pipeline role from Part I: a workflow that looked operationally normal, accumulated broad authority through reasonable changes, and quietly became a cross-environment control surface. Deployment is not the only important workflow, but it is a useful first case because it concentrates the book's central concerns: authority, speed, trust, automation, and consequence.

The same build sequence applies to other high-impact workflows: break-glass debugging, bulk data export, privileged identity issuance, vendor integration, and external actions taken by agents. Each one asks the same design question: which unsafe trajectory should no longer be reachable through normal work?

Most enterprises already operate pieces of this architecture. Cloud identity systems mint temporary credentials. Policy-as-code engines evaluate conditions. Workload identity frameworks attest services. CI/CD platforms track workflow state. Internal developer platforms expose supported actions. Graph systems model relationships and reachability. The problem is that these pieces are usually deployed as separate controls rather than composed into a system that governs trajectories.

This chapter avoids prescribing a single product stack. A transition model may reside in a workflow engine, a deployment controller, or an incident-management system. An authority foundry may be implemented through cloud federation, a secrets broker, workload identity, or JIT access. Curated actions may appear as platform APIs, service-catalog actions, or paved road workflows. Sequence integrity may be enforced through policy-as-code, admission control, or deployment gates. Collapse may be implemented through token leases, session brokers, or workflow-state-triggered revocation.

Technology will vary. The design responsibility doesn't.

Start with a Transition Model for a High-Impact Workflow

Trajectory governance begins by changing the unit of design. Most security models start by asking the following:

- Who can deploy?
- Who can assume a role?
- Who can export data?
- Which API calls are permitted?

This framing is familiar but narrow, and it often breaks at speed. Trajectory governance, however, starts one step earlier with a different question: Is this a valid next move from the current state of the workflow?

This question sounds like a subtle change in phrasing, but it really isn't. It changes the control surface entirely. The system is no longer being asked whether an action is ever allowed in the abstract. It is being asked whether that action is reachable from here and now, given what has already happened. That is how you begin governing trajectories rather than only permission.

Choose One Workflow Where Failure Is Real

Don't try to model the whole environment first. Pick one workflow where the consequences are meaningful, and the pressure is constant.

In most organizations, the first candidates are easy to name:

- Production deployments
- Break-glass debugging
- Bulk data movement
- Privileged identity issuance
- External actions, such as sending messages, issuing refunds, or changing customer entitlements

The production deployment path is a strong place to start, because it already concentrates several of the book's core concerns: It's high authority, it's frequent, it runs under delivery pressure, and it often becomes a quiet source of trust inflation through shared pipeline roles, inherited secrets, and cross-environment reuse. A deployment path isn't just "moving code." It's a control-plane workflow that can alter runtime reality across important environments and makes it ideal for a first deletion.

Name States That Reflect Readiness, Not Infrastructure Trivia

A useful transition model names the states about which the control plane can reason. Not "cluster exists," but "this artifact has been validated." Not "user has admin," but the "incident declared" and "debug session minted." States should reflect *what stage of work you are in,* because it's what determines which moves should become reachable.

For a deployment path, a workable first model looks like the following:

- Change proposed
- Change reviewed
- Artifact built
- Artifact validated
- Deploy authorized
- Deploy executed
- Deploy complete

This state doesn't have to be the perfect state machine on day one. It only has to be concrete enough that the system can tell the difference between legitimate progression and unsafe trajectory.

In some organizations, you may add specific intermediate states, such as security exception granted, rollback pending, or incident-declared recovery mode. That is fine. The key isn't formal elegance. The key is about making the workflow legible enough that the control plane can reason about what should be reachable next. The same design pattern applies to other scenarios as well.

A break-glass workflow might include:

incident declared → session justified → session minted → session active → incident closed → session collapsed

A data export workflow might include:

request submitted → purpose validated → export approved → export executed → export window closed

An agent external-action workflow might include:

gather → draft → approve → execute → terminal collapse

These aren't documentation exercises. They are the minimum structure required to make a trajectory governable.

Delete One Unsafe Trajectory by Naming It

The most operational move you can make in this entire book is to *explicitly name what must become unreachable.* Examples are often simple:

- Deploy to production *without* a validated artifact.
- Mint privileged access *without* a workflow context.
- Export sensitive data *without* a purpose-bound pathway and terminal state.
- Allow an agent to execute an external action *without* an approval state.

These aren't "policy reminders." They are *reachability deletions.* Your goal isn't to detect these transitions faster. Your goal is to engineer the system so that those moves don't exist in the reachability graph.

Treat the Model Like an Engineering Artifact

A transition model that lives only in a slide or a Wiki page isn't governance. It is an aspiration.

The output should look like something that an engineering team can build against:

- A state diagram (even a rough one)
- A list of valid transitions
- A list of transitions that must become unreachable
- A few plain-language "this must never be possible" tests

For the deployment example, the tests might read as such:

- Production deploy is unreachable until artifact validation is present.
- A deployment token cannot target a different environment than the one authorized in the workflow state.
- The system that executes deployments cannot modify the policies that govern deployment authorization.
- Rollback authority cannot persist past the recovery workflow that minted it.

Once these elements can be stated concretely, the problem shifts from describing transitions to governing authority: identifying where the power to perform them originates and ensuring that it can be established only through valid pathways. That's the authority layer.

Build the Authority Layer: An Authority Foundry, Not Free-Form Power

In most organizations, authority is easy to create and hard to unwind:

- A team needs access, so a role is widened.
- A pipeline needs permissions, so a service account becomes broader.
- A tool needs integration, so a token persists indefinitely.

It is how *trust inflation* becomes normal: Trust expands through convenience and reuse, crossing boundaries that were intentionally designed to remain separate. The expansion often looks rational locally, for example, "just add this permission" or "just reuse that role," and it becomes dangerous globally because it accumulates into load-bearing authority that no one can easily reduce later.

Contraction is expensive, political, and late. Structural constraints require an architectural reversal: Certain classes of authority must only originate through a constrained "foundry" that mints bounded, purpose-bound, short-lived power as part of a valid workflow. This is not an approval queue. It is a control-plane primitive: The one place allowed to mint certain kinds of authority, under explicit prerequisites, with a terminal state.

The point is simple: Free-form authority authoring must stop being the default operating model.

What to Build at This Layer

At minimum, the authority layer needs three things:

1. **Workflow identity that the control plane can trust**
   Not mind-reading; just explicit context like incident identifications (IDs), deployment job identities, approval artifacts, and ticket references
2. **An issuer/broker that mints bounded authority**
   The interlock—the only system allowed to mint certain privileges, tied to workflow context, scoped, time-bounded, with a terminal state

3. **A small capability catalog + compiler**
   A curated surface area for intent ("what I need to do") and a compiler that produces bounded authority ("what power exists"), with versioning and change control at the capability level

For most organizations, two deliverables usually come first, because they remove the most recurring sources of risk and friction.

Deliverable 1: Credential-less Work as a Default Workflow

In many environments, developers, pipelines, and tools still "carry keys." Even when those keys are technically short-lived, they still are copied into tools, cached on laptops, embedded into CI systems, reused across tasks, and remembered long after the intent that justified them is gone.

Trajectory governance flips the abstraction. For instance, take the deployment workflow. In the old model, the pipeline holds a standing role broad enough to deploy across several environments, fetch secrets, and modify release state. The system moves fast, but it does so by carrying durable power.

In a foundry model, the pipeline doesn't begin with broad-standing power. Instead, it takes these steps:

1. An operator expresses intent ("debug this service," "run this migration," or "inspect these logs").
2. The system confirms that the workflow is in a state where deployment is valid.
3. The foundry mints short-lived deploy authority scoped to that service, artifact, environment, and time window.
4. Once the workflow reaches its terminal state, that authority collapses automatically.

The shift is: Access becomes a state transition, not a possession.

That same move applies outside deployment:

- A break-glass session isn't a standing administration role. It is a workflow-bound session minted after incident declaration, scoped to the service or environment under investigation, and collapsed when the incident closes.
- A data export isn't "whoever has the right token can move the data." It is a purpose-bound pathway minted for one dataset, one reason, one sink, or one time window.

Credential-less work slows trust inflation because it stops producing the leftover authority that later gets reused. It doesn't merely make credentials shorter-lived. It changes what "having access" means.

Deliverable 2: Capabilities as the Default Way to Request Authority

The second-recurring risk is free-form authority authoring: letting engineers express machine authority directly through raw IAM/RBAC/policy documents, copied templates, and bespoke grants. Even with reviews, the pattern is predictable:

- Local pressure expands authority.
- Authority persists longer than intent.
- Roles accrete through reuse.
- Broad identities become bridges across boundaries.

The structural alternative is to change the unit of expression.

Instead of authoring authority directly, teams request *capabilities*, which are intent-level primitives such as:

- read dataset X
- deploy service Y
- rotate secret Z
- publish event to topic T

The platform then compiles that request into minimal authority using vetted patterns and bounded scope.

A simple litmus test captures the shift:

- If someone can unblock themselves by writing a new permission document, you're still governing by memory and review.
- If they unblock themselves by selecting a supported capability and the platform mints the smallest sufficient authority as output, you're governing by structure.

It's also how you prevent trust inflation from becoming your default operating mode. Free-form authority authoring is a trust inflation machine: It makes expansion cheap and contraction expensive. A capability model reverses that: Expansion becomes deliberate and visible because it requires extending a constrained vocabulary rather than sprinkling privileges across the environment. AI can help here, but only if it's used in the right role:

AI is the interpreter, not the foundry.

AI can translate messy requests into the right capability, propose the smallest sufficient scope, generate the request, and explain trade-offs. But it shouldn't be allowed to mint arbitrary authority directly. Otherwise, you'll get to the old world, only faster. (This also tracks according to the first principles listed in Appendix A).

Authority solves origination. But trajectories will still fail when the interface is raw, and the system allows unsafe sequences. That's the next layer.

Build the Interface and Progression Layer: Curated Actions and Sequence Integrity

If the default interface for high-impact work is raw control-plane power (Terraform/YAML/API calls), you're forcing every developer and every agent to be a security architect during routine execution. That isn't autonomy-ready, though, and it isn't a scalable way to govern decision space.

Trajectory governance requires two complementary constructions:

1. **A steering wheel:** Curated actions that reflect intent
2. **Rails:** Progression rules and terminal states that enforce a valid sequence (i.e., sequence integrity)

You need both, as a steering wheel without rails still allows unsafe sequences. Rails without a steering wheel still force people into the engine room. It's also why curated actions and sequence integrity shouldn't be mistaken for a more elaborate allowlist. The distinction is about governing valid progression, not just permitted steps. (See Appendix B for more information.)

Steering Wheel: Expose Intent-Level Actions, Not Raw Capability

Curated actions are governance primitives. They reduce what can be expressed in the first place. For the deployment workflow, a curated action might look like the following:

deploy(service=payments, artifact=validated_digest, environment=prod)

That is a radically different security surface from "this identity can push arbitrary configuration and artifacts across environments." Other recurring actions may look like the following:

- rollback(service=payments, deploy_id=123, incident=INC-45)

- read_logs(service=X, window=30m)
- export(dataset=Z, purpose=P, sink=S, budget=B)
- break_glass(service=S, incident=I)
- send_external(template=T, audience=A, approval=R)

These actions aren't merely easier interfaces. They're narrower interfaces that define the ceiling of what is expressible.

A raw API says, "Here is power; use it correctly." A curated action says, "Here is the bounded move the system is willing to support." Curated actions reduce the pressure that drives people toward inherited roles, bespoke scripts, and free-form trust expansion.

In the deployment case, a curated action means teams aren't given a single general pipeline role and asked to reason about everything it might do. They are actually given one supported move. The blast radius is bounded by the interface itself.

Rails: Make Unsafe Sequences Unreachable

The most damaging incidents are rarely a single forbidden action. They are allowed actions taken in the wrong order. So, the control plane must enforce progression, not just permission. For the deployment path, it means rules like the following:

- Deployment is unreachable until validation is committed.
- Production deployment is unreachable until a review state exists.
- The deploy token is valid only for the specific service, environment, artifact, and time window that's tied to the workflow.
- Rollback authority is reachable only from failed-deploy or incident states.

This is where "trajectory" becomes enforceable. The system stops asking only: Is this identity allowed to do *x*? and it starts asking: Is *x* a valid next step from where we are right now? This second question is what collapses whole classes of autonomy failures. It's also a direct counter to trust inflation. When trust expansion isn't a side effect of execution, the system stops widening boundaries silently through reuse.

The same rail pattern applies to the other examples.

For break-glass debugging:

- A privileged session is unreachable until an incident state exists.
- The session is bounded to one environment or service.

- Terminal collapse is automatic when the incident closes.

For data export:

- The export is unreachable until its purpose and sink are declared.
- The export path is bounded by its destination and budget.
- The authority collapses when the transfer window closes.

For agent external action:

- Gather may be reachable.
- Draft may be reachable.
- Execute is unreachable until approval state exists.
- Authority collapses once the approved action completes or expires.

What Changes Once This Layer Exists

When curated actions and sequence integrity are real, a surprising number of recurring security campaigns start disappearing as standalone problems:

- "Temporary" access stops becoming permanent because workflows terminate by design.
- Data movement becomes brokered and purpose-bound, instead of being ambient and ad hoc.
- Pipelines stop being able to widen their own authority.
- Sensitive external actions stop being reachable through generic interfaces.
- Agents stop inheriting broad power simply because the surrounding human identity happened to have it.

At this point, the system is technically governable under speed.

However, technical governability isn't enough. If teams route around the whole stack on Friday afternoon, during an outage, or when the deadline is immovable, then the constraints aren't yet real organizationally.

Ship the Paved Road: Make Constraints Propagate Under Pressure

By now, the pattern should feel clear:

- A transition model without adoption is just a diagram.
- An authority foundry without adoption is a queue.
- Curated actions nobody relies upon are just well-written wrappers.

Structural constraints only become real inside an organization when they *propagate*—when they become the default way work happens, even on a Friday afternoon, even during an incident, or even under machine speed.

Such action is the job of the paved road. While a curated action is a supported safe way to perform a single high-impact task, a paved road is the end-to-end default workflow that makes those curated actions the normal way work is accomplished.

The "paved road" metaphor isn't accidental. Netflix popularized the idea of a paved road as a supported path: It's a standardized, well-integrated set of tooling and commitments that teams can choose to follow (and when they do, they benefit from stronger support), without making it mandatory. Spotify describes a closely related model through "golden paths," emphasizing onboarding, consistency, and the discipline of improving the path itself.

A paved road isn't a secure template that you can optionally adopt. It is the primary interface through which high-impact work is done. It is how structural interlocks become ordinary behavior without requiring constant vigilance. A paved road is the productized form of everything built in this chapter: It's the transition model, foundry, curated interface, progression rules, and collapse logic all packaged as the supported path teams actually choose.

For the deployment workflow, the paved road shouldn't be a pile of guidance. It should be a usable system.

A team onboarding a new service should inherit the following by default:

- The supported deployment template
- Validated-artifact requirements
- Purpose-bound deployment identities minted by the foundry
- Environment separation
- Rollback behavior
- Auditability
- Progression rules are already wired into the flow

In a strong paved-road model, teams don't begin by inventing deployment and then ask whether it's secure enough. They begin by choosing the supported deployment path. It's the same organizational shift that Chapter 6 described, now translated into concrete engineering. In a trajectory-governed system, secure properties must be inherited, not reapplied manually for every new service.

"Secure by default" has been repeated for years. The upgrade here is structural. If each new workload still requires a bespoke review to reach baseline safety, then the organization hasn't built structural constraints. It has built a repeatable ritual.

Inheritance is what turns one good design into an organizational operating model. In the deployment example, inheritance means a new service that uses the paved road automatically gets bounded deploy authority, validated artifact progression, narrow blast radius, and reliable collapse. It's what makes the model scale.

Subsidize the Safe Path; Tax Decision-Space Expansion

A paved road doesn't win merely because it exists. It wins because it is economically better.

The rule is simple: Subsidize the safe path; tax decision-space expansion.

For deployment, that means the supported path should be fast, reliable, well-documented, well-integrated, and easier than custom alternatives. The off-road path shouldn't always be impossible, but it should be more deliberate, more visible, and less convenient when it expands authority or blast radius. So …

- using the supported deployment action should be one normal move,
- asking for a broad cross-environment standing role should be slower and more exceptional,
- bypassing artifact validation shouldn't feel like "the quick route," and
- custom generic pipeline power shouldn't be easier than the supported path.

The point isn't punishment. The point is making the system's intended shape visible in a day-to-day experience. Here's how this can look for the other examples we discussed earlier:

- A break-glass paved road should be faster than informal privilege borrowing.
- A supported export path should be easier than stitching an ad hoc data move.
- A bounded external-action workflow for agents should be easier than handing the agent a broad inherited token.

That is how constraint design survives real operational pressure. In addition, it's the only way trajectory governance scales beyond a small number of heroic engineers who understand the system.

Adoption Is the Control

Most security programs treat adoption as a change-management problem: "We need teams to comply." That framing is already a warning sign. It implies the secure path is something teams must be convinced to choose.

In a trajectory-governed system, adoption is a design property. The secure path wins because it is ...

- *faster* than doing it yourself,
- *more reliable* than doing it yourself,
- *more complete* than doing it yourself, and
- *the only path that works immediately* for high-impact work.

When those conditions hold, you don't need to persuade the organization. You do not need to police the path. You don't need to run a permanent exception economy. You simply ship the paved road, and the organization drives on it. It is what *institutionalizing* structural constraints means.

What a Paved Road Actually Contains

A paved road consists of a small number of building blocks that teams assemble repeatedly. The mistake is to think those building blocks are "security controls." They are workflows and primitives that encode intent, mint authority safely, and constrain transitions by construction.

In practice, a category isn't really "done" because it's documented. It is done when ...

- the boundary is enforceable in the control plane,
- the curated action exists and covers real work,
- collapse is automatic and reliable, and
- teams actually use the supported path under pressure.

That last bullet prevents self-deception. If the organization still relies on off-road execution for the common or urgent version of the workflow, then the control is incomplete, no matter how elegant the model looks.

The building blocks the paved road consists of are usually some combination of the following:

- **Opinionated starting points:** Blueprints, modules, and scaffolds (call them what you want, but they aren't "samples") are supported products. They "bake in" the baseline constraints that should never be reinvented per team.

- **Curated actions for recurring high-impact work:** A paved road wins by offering a small set of moves that cover most real work.
- **An authority foundry that is mostly invisible:** If the paved road is successful, most users never think about the issuer. They simply experience that the system grants the right authority at the right moment and then collapses it when the work is done.

Where These Primitives Live in Real Architectures

Table 8.1: Trajectory-governance primitives and technology anchors

Trajectory Governance Primitive	What It Does	Current Technology Anchors
Transition Model	Represents workflow state and valid next moves.	Workflow engines, state machines, deployment orchestrators, change-management systems, CI/CD metadata, incident-management state, internal developer platforms.
Authority Foundry	Mints bounded authority only inside valid workflow context.	Cloud STS/IAM federation, privileged-access brokers, secrets brokers, workload identity systems, SPIFFE/SPIRE, Vault-style leased credentials, JIT access systems.
Curated Actions	Exposes intent-level moves instead of raw control-plane power.	Internal developer portals, platform APIs, Backstage-style golden paths, Terraform modules, service catalogs, API gateways, deployment templates.
Sequence Integrity / Stateful Progression	Ensures actions are valid only from the right workflow state.	Policy-as-code engines, admission controllers, CI/CD gates, workflow engines, OPA/Rego, Cedar, Kubernetes admission control, deployment controllers.

Trajectory Governance Primitive	What It Does	Current Technology Anchors
Automatic Collapse	Ensures authority ends when the mission ends.	Time-bound credentials, token leases, incident-state-bound access, certificate/SVID rotation, JIT/PAM session brokers, workflow-triggered revocation.
Decision-Space Map	Shows how authority, reachability, workflows, and trust compose.	Graph databases, knowledge graphs, identity graph analytics, attack-path tools, OpenFGA/Zanzibar-style relationship models, cloud-identity graph models.
Paved Road	Packages the primitives into the default workflow teams actually use.	Internal developer platforms, golden paths, service templates, CI/CD platforms, platform engineering portals, self-service workflows.

None of these technologies, by itself, is trajectory governance. OPA can express policy, but the organization still has to decide which transition should be valid. Cedar can support authorization decisions, but the workflow still has to define what authority means in context. SPIFFE/SPIRE can provide strongly attested workload identity, but identity alone does not guarantee bounded purpose, valid progression, or collapse. Graph databases can model decision space, but the model only matters if it changes what the platform allows. The primitive is the design responsibility. The tool is the implementation surface.

The Exception Economy Is a Product Signal, Not a Compliance Failure

Every organization has exceptions. What matters is what you do with those exceptions. In a posture-based world, exceptions are paperwork. They can accumulate and become debt. In a trajectory-governed world, exceptions are data.

When teams bypass the paved road, it usually means one of three things:

1. The road is missing a needed curated action.
2. The road is too slow or unreliable.
3. The road's constraints are misaligned with how real work happens.

Treating bypasses as "noncompliance" is the fastest way to turn your paved road into a ghost town. Treating bypasses as product feedback is how paved roads improve.

For the deployment example, the responses that scale are engineering responses:

- Ship the missing deploy or rollback path.
- Widen the coverage of the supported flow.
- Fix the collapse so that teams aren't stranded mid-incident.
- Remove the convenience of the off-road route, which keeps recreating trust inflation.

Once structural constraints are encoded, authority is minted safely, interfaces are constrained, and progression is governed, the security problem changes shape. You are no longer primarily asking:

- Are we compliant?
- Did someone do something unsafe?
- Can we detect violations faster?

You are now asking:

- Are teams staying on the supported path?
- Where is pressure building at the boundary of the decision space?
- Which constraints are too permissive, and which ones are too constraining?
- Which unsafe trajectories are still reachable because the road is incomplete?
- When something goes wrong, what does it reveal about the transition model?

These are no longer design questions.

Governing the Governor: Securing the Paved Roads

A paved road concentrates authority. That is its purpose. It centralizes the transition model, the authority foundry, the curated action, the progression rule, and the collapse condition into a path that teams can safely use under pressure. But this also means the paved road itself becomes a high-authority control-plane asset. That creates a new responsibility. The system that governs authority must itself be governed.

If an adversarial engineer, compromised pipeline, malicious insider, supplier-controlled tool, or autonomous agent can modify the foundry rules, policy bundles, workflow states, or paved-road configuration, then they don't need to bypass the paved roads. They can redefine it. This will be the failure mode of the ungoverned governor.

Trajectory governance does not solve this by trusting the foundry more. It solves it by applying the same structural discipline to the foundry's own configuration path. The authority foundry, policy engine, curated-action catalog, paved-road templates, workflow states, and collapse rules must be treated as Tier-0 governance assets. Their configuration paths are among the most consequential decision paths in the enterprise. This is not a new control level. It is structural controls applied to the governance plane itself.

The first requirement is an immutable source of truth. Foundry rules, policy bundles, workflow-state definitions, paved-road modules, and curated-action catalogs should be stored in versioned, reviewable repositories. Runtime systems should consume signed and released versions of those definitions. They should not depend on direct console edits, unmanaged configuration changes, or ad hoc administrator updates.

The second requirement is multi-party verification. Changes to the paved roads should require stronger controls than ordinary application changes. A new foundry rule or policy bundle should require separation of author and approver, cryptographic signing, provenance, and a governed promotion path. The point is not to create a ceremony. The point is to recognize that changing the governor changes what futures the enterprise can reach.

The third requirement is stateful progression. Governor changes should move through explicit states:

Proposed → Reviewed → Tested → Attested → Released → Activated → Monitored

A policy bundle should not become active merely because a file has changed. The system should verify provenance, test the change against known safety claims, and promote it through a controlled workflow. The same design logic used for production deployment applies to the governance plane itself.

The fourth requirement is separation between execution and governance. The system being governed should not be able to rewrite the rules that govern it. A CI/CD pipeline should not be able to modify the policy that decides whether the pipeline can deploy. An agent should not be able to alter the broker that mints its tool authority. A workload should not be able to change the identity rule that grants its workload identity. A break-glass user should not be able to permanently weaken the break-glass pathway.

The fifth requirement is trajectory regression testing. Every governor change should be tested against the unsafe paths the organization has already deleted. Production deployment without validated artifact provenance must remain unreachable. Privileged debugging without an active incident must remain unreachable. External agent action without approval state must remain unreachable. Authority must still collapse when the workflow reaches its terminal state. A pipeline must still be unable to widen its own authority.

These are not ordinary unit tests. They are tests of the organization's structural safety claims.

Table 8.2: Governor surfaces, failure modes, structural protections

Governor Surface	Failure Mode	Structural Protection
Foundry rules	A policy change allows broader authority to be minted.	Signed policy bundles, multi-party approval, provenance, and regression tests for forbidden trajectories.
Paved-road templates	A template quietly reintroduces broad roles or long-lived credentials.	Immutable source repositories, reviewed releases, artifact attestation, and deprecation of unsigned templates.
Curated-action catalog	A curated action becomes a disguised admin action.	Capability design rules, scope budgets, versioning, and approval for new high-impact actions.

Governor Surface	Failure Mode	Structural Protection
Workflow-state model	A state transition is added that skips validation, approval, or collapse.	Explicit transition diagrams, stateful promotion, test cases for invalid transitions, and monitored activation.
Policy engine / broker configuration	Runtime configuration is changed directly outside the governed path.	No direct console edits, signed release channels, separate admin authority, and tamper-evident audit trails.
Break-glass path for the governor	Emergency access becomes a permanent bypass around the paved roads.	Incident-bound access, multi-party activation, short-lived authority, automatic collapse, and post-use review.

The key principle is: The paved road can't be allowed to pave over its own guardrails.

The authority foundry must not mint unbounded authority for itself. The deployment pipeline must not change the policy that decides whether it may deploy. An agent must not alter the broker that constrains its tools. A break-glass administrator must not convert emergency authority into a permanent bypass. Across all of these surfaces, the same rule holds: the system being governed must not be able to weaken the path that governs it.

This is where trajectory governance meets software supply-chain discipline. Foundry rules, policy bundles, workflow-state definitions, paved-road modules, and curated-action catalogs should be treated as high-integrity artifacts: versioned, signed, attested, promoted through known states, and tested against safety invariants before activation. A trusted path is not trustworthy because the organization declares it is trusted. It is trustworthy because its own change path is constrained.

A trajectory-governed organization must be able to prove two things. First, unsafe business workflows are constrained by design. Second, the governance paths that constrain them can't be casually rewritten under pressure. Without the second proof, the first one is temporary.

The governor must be governed.

CHAPTER 9

Operationalizing Intent: Operating in Decision Space

The on-call pager still goes off.

Deployments still run. Incidents still happen. Engineers still need access. Pipelines still request credentials. Agents and automations still call tools, update systems, and move faster than teams can keep up with in real time. The work does not become quiet because the safer path has been built. The question is whether that path holds when real work begins.

That is the work of this chapter. Once high-impact workflows have supported paths, scoped authority, curated actions, progression rules, and automatic cleanup, security operations get a different job. It is no longer only about watching alerts, logs, suspicious events, and policy violations after they appear. It is watching the moments when access is granted, credentials are issued, privileged actions are taken, exceptions are requested, bypasses occur, and temporary authority fails to end.

On a normal day, those signals may look routine. On a bad day, they matter. A service is failing. Customer impact is rising. A production deployment needs a rollback. An engineer needs a diagnostic action that the supported path almost covers, but not quite. These are the moments where old habits return. Someone asks for a longer session, a broader role, a raw console, a standing token, or one temporary exception that will be cleaned up later.

Security operations has to read those moments correctly. A bypass may mean the supported path is missing a real use case. A lingering token may mean that cleanup is still too manual. Repeated denial may mean the workflow is blocking legitimate work. A raw console fallback may mean the curated action is too narrow. Not every exception is misconduct. Not every denial is a success. The job is to understand whether the system is holding, where it is bending, and what needs to change

This changes what operators watch, how they triage, and what done means after an incident. The question is no longer only whether something unusual happened. It is whether the wrong authority was issued, the wrong path was used, a temporary exception became normal, or access remained after the work was over. Security operations still responds to incidents. But in this model, it also helps reduce the chance that the same unsafe path remains available next time.

What Changes When You Operate This Way

Most security operations today begin with an event: an unusual login, a suspicious API call, a burst of access, or a policy change. The event becomes the story: investigate, contain, restore, and move on.

In a trajectory-governed system, the event is rarely the story. It is a symptom. The story is the pathway that made the event possible:

- What intent was expressed
- What authority was minted in response
- Which paved-road–curated actions were used (or bypassed)
- Whether the system returned to a safe terminal state when the work was done

This isn't a small change in tooling. It's a change in what your team considers "operations" in the first place. Most organizations will do this in a hybrid environment for years. Some workflows will be trajectory-governed, while others will still posture-based. This chapter focuses on operating the paved roads. Chapter 11 covers how to expand them safely.

You will still need logs, detections, and response. But they will stop being your primary lens. The highest-leverage operational question will no longer be "What happened?" It will be "What boundary bent, and what did that make reachable?" That shift changes where your attention goes.

Instead of trying to interpret every event equally, the team begins focusing on a smaller set of higher-leverage signals: entries into privileged pathways, high-impact transitions, off-road execution, authority that fails to collapse, and repeated pressure against the same structural boundary.

This is why the model isn't simply "better detection." It is a different object of operations. The team is no longer primarily watching for bad behavior in the exhaust of the system. It is watching the places where the system's decision space is being widened, stressed, or bypassed.

Response changes, too. In a posture-based model, response usually has one deliverable:

- Restore safe operation.
- Revoke access.
- Roll back the change.
- Contain the incident.
- Get the environment back to an acceptable state.

In a trajectory-governed model, response has *two* deliverables:

- First deliverable of restoration. You contain, recover, rotate, patch, and restore service. That work remains essential.
- Trajectory governance adds a second deliverable that is structural: Make the same unsafe trajectory unreachable next time.

A posture-era ticket often sounds like this:

- Reduce permissions
- Tighten the policy
- Remind the team
- Add a detection

A trajectory-governed outcome sounds different:

- This authority can no longer originate that way.
- This transition is no longer valid without workflow context.
- This privileged path now collapses automatically.

- This off-road route no longer exists as a normal option.

That is a different definition of "done."

An incident isn't truly closed because the immediate damage has been contained. It is closed when the system has learned structurally and when the path that produced the failure has been narrowed, bounded, or deleted. It's why operations start to look less like event processing and more like architectural debugging. The team is no longer only asking how to restore the environment. It is asking what the environment just revealed about the incompleteness of its constraint design.

Once you start operating at the boundary, you'll see a pattern that event-centric security often misses until it is too late: The boundary doesn't usually break all at once. It drifts, and t*hat drift is the leading indicator of failure in high-speed environments.*

Before we discuss drift, however, we need to make boundary operations concrete. We need to define what it looks like on a bad day.

What You Watch: Boundary Operations Under Pressure

The fastest way to understand this operating model is to place it in the most common high-pressure scenario in modern environments: an outage. In this outage, a service is failing, a customer impact is rising. The on-call engineer needs answers now.

In a posture-based world, this is where the system quietly teaches the wrong lesson: that speed comes from improvisation. Broad access is granted "temporarily." Work is done. Cleanup is deferred. The next incident arrives before cleanup happens. Over time, the environment becomes a collection of half-revoked privileges and ad hoc pathways created under stress and never fully dismantled.

Trajectory governance doesn't try to prevent emergencies. It makes emergencies governable.

A Bad Day, Redesigned

In a trajectory-governed environment, the break-glass moment isn't a social workaround. It is a defined transition:

1. An incident is declared (or referenced).

2. A break-glass pathway is entered.
3. Bounded authority is minted with explicit scope and duration.
4. Operators work through paved-road *curated actions* that provide what they actually need.
5. The pathway collapses when the incident ends.

This changes what operations watches. The important question is no longer: What commands did the engineer run? That is noise unless you already know where to look. The real questions become:

- Was this break-glass entry expected for this service and environment?
- What authority was minted and for how broad, how long, and for what declared purpose?
- Did the pathway terminate cleanly, or did it linger?
- Did the operator use the supported curated actions, or did they bypass into raw control-plane capability?

Those are boundary questions. They are legible, auditable, and actionable. Most importantly, you can answer them quickly even when the system is moving fast, because the signal is produced at the moment authority is created and not after the damage is done.

What You Instrument: Intent-to-Authority Conversion, Not Just Activity

Most organizations already log activity, including cloud audit trails, API calls, deploy events, and access records. They are the table stakes. The upgrade here is that you log the conversion:

intent → pathway → minted authority → transition → terminal state

You don't need perfect semantics. You need structured linkage.

In practice, every high-impact paved-road pathway should emit a minimal boundary record that answers the following:

- Who or what invoked the pathway (the user, workload, agent, or pipeline identity)
- Why (including the declared intent category plus a reference such as incident ID, change ID, or workflow ID)
- Which pathway was used (a supported curated action or override route)
- What authority was minted (a capability, scope, duration, or environment)
- What transition occurred (deploy, export, exposure, elevation

- Whether it terminated (terminal state reached, and when)

This record isn't part of bureaucracy. This record is your new operational requirement. Without it, an incident response team will still be forced to reconstruct intent from fragmented information across Slack messages, tickets, and log archaeology, and you will end up right back in the old world of faster forensics, with the same structural risk.

What the Boundary Dashboard Looks Like

If you're used to dashboards full of detections, the boundary dashboard will feel almost too small. That's the point, though. A boundary dashboard isn't a wall of red. It is a small number of trend lines and queues that represent how your constraint system behaves under real usage.

At a minimum, the boundary dashboard should show and answer its respective questions:

- **Privileged pathway entries:** They should show break-glass sessions and privileged workflows and answer: *How often? By whom? How long? Are terminal states reached reliably?*
- **High-impact transitions:** They should show deployment-to-production, exposure changes, and bulk data movement, and answer: *How often? Through which curated actions? With what authority shape?*
- **Overrides and bypasses:** They should answer: *Which teams? Which systems? Which curated actions are missing, too slow, or misfit for real work?*
- **Denied transitions:** They should answer: *Are these probes, one-off mistakes, or pressure signals that the paved road is incomplete?*

Notice what's missing: a hundred "suspicious event" categories.

You still need detection and response. But your primary operational lens shifts to operate the boundary, not the exhaust.

How Triage Changes

When operations are event-centric, runbooks tend to end with the restoration step. When operations is boundary-centric, runbooks add another layer: Improve the constraint system.

A boundary-oriented triage loop usually looks like the following:

- Classify the interaction: privileged pathway, high-impact transition, override, or denied transition
- Validate it against declared intent: Is there a legitimate reference, and was the right pathway used?

- Check the scope and terminal behavior: Was a minimum authority minted, and did it collapse cleanly?
- Decide whether the paved road or the actor is the issue.
- If people bypassed, ask whether the paved road is missing a curated action, is too slow, or is misaligned with real work.
- If someone forced an override, ask whether it is truly rare or becoming routine.

The characteristics change the feel of operations. Even when the outcome is "everything is fine," you have learned something valuable: The system held under pressure, and you have evidence that the boundary worked as designed.

When the outcome is "something is wrong," you have a precise diagnostic target: the point where intent became authority, where the wrong pathway was used, where the scope was too broad, or where collapse failed.

This is why boundary operations feel different from traditional incident response. It isn't only about restoring the service. It's about understanding whether the emergency path itself remains governable under stress. Once you start operating this way, you'll see the next pattern quickly: The boundary usually doesn't fail all at once. It bends, stretches, and drifts.

Drift usually begins as a reasonable step. The paved road works for the common case, but the next incident is slightly different. The diagnostic action exposes most of what the operator needs, but one dependent system is missing. The session duration works for routine debugging, but an outage needs a longer window. The collapse condition works cleanly for one service, but another service remains under investigation. None of these moments feels like a breakdown in governance. They feel like operational reality is pressing against the boundary.

That is why drift is easy to misread. If every pressure signal is treated as noncompliance, teams will learn to hide pressure. If every pressure signal is accommodated by widening authority, the organization quietly rebuilds the old model inside the new one. The operating task is to distinguish friction that reveals a product gap from pressure that is trying to reopen an unsafe trajectory.

A trajectory-governed system must remain teachable without becoming permissive. When a team asks for a longer session, the answer is not automatically no. The better question is why the session length is insufficient for legitimate work. When a team bypasses a curated action, the answer is not automatic enforcement. The better question is whether the curated action actually covers the operational path the team had to take. When a workflow fails to collapse cleanly, the answer is not merely a cleanup ticket. The better question is which terminal condition the system failed to represent.

The boundary is not a static wall. It is an operating surface. It must hold under pressure, but it also has to reveal where its current shape no longer fits reality. Drift is the signal that this tension has become visible. It shows where the constrained system, as designed, and the organization, as operated, are beginning to diverge.

The danger is not that teams ask for exceptions. The danger is that the same exception becomes familiar. A request becomes a pattern. A pattern becomes a shortcut. A shortcut becomes a dependency. A dependency becomes infrastructure. At that point, the organization is no longer making an exception to the governed path. It has created a second path, usually broader, less visible, and less likely to collapse.

Detecting drift early prevents that second path from hardening into the new operating model.

What Drift Looks Like Before It Becomes an Incident: Detecting Pressure Early

Decision-space drift isn't the same as configuration drift.

- Configuration drift is the state change.
- Decision-space drift is the space of reachable outcomes expanded.

Most important failures in a trajectory-governed system don't begin as obvious incidents. They begin as quiet expansions in what the system can now do: They expand to broader authority shapes, longer-lived elevated states, repeated bypasses, more permissive pathways, and trust edges that outlive the reasons they were created.

In hybrid environments, drift is often imported. Teams route around the paved road using legacy roles, ad hoc tokens, raw control-plane APIs, and inherited cross-boundary trust. Left unchecked, the posture-based side becomes a trust inflation engine that eventually spills into the governed side.

You don't detect drift by looking for "bad events." You detect drift by watching for pressure against the boundary. That's the real shift in operating model. Drift isn't simply a misconfiguration that appears over time. It is evidence that the constrained system and the real world are diverging.

Drift Is What Happens When Exceptions Become a Workflow

The most common drift pattern is: The paved road exists, but it doesn't fully cover reality.

A team hits a situation that a paved road doesn't handle. They go off-road to get their work done. An override is granted. A one-off capability is minted. A "temporary" broad permission is issued. An emergency pathway stays open longer than intended. Each decision is defensible in isolation. All of them are rational under pressure. But the aggregate effect is that the system quietly contains far more reachable futures than anyone intended.

This is also where trust inflation becomes visible operationally. Trust inflation isn't only a design flaw; it's a trend line. It shows up as repeated reuse of powerful pathways, expanding cross-boundary trust, and accumulated authority that outlives the reasons it was granted. In this sense, drift isn't a separate problem from trust inflation. It is how trust inflation looks while it's still becoming normal.

Let's look at an example where credential-less work starts to decay. Credential-less workflows are a perfect example, because drift is easy to recognize:

You ship an access pathway that issues purpose-bound, time-bounded authority tied to an incident. It works, and your teams adopt it. Everything looks healthier.

Then pressure appears:

- A team asks for a longer session because "incidents take hours."
- Another team asks for a standing token because "on-call gets interrupted."
- A pipeline needs to call an internal service, and someone proposes "Just store the token in CI so it doesn't break again."
- A third team bypasses the issuer entirely, because a niche tool doesn't integrate with the paved road.

None of these requests are malicious. They are rational optimizations.

But if you approve them repeatedly, you don't just solve friction. You change what is reachable:

- Long-lived authority reenters the system.

- Tokens become portable artifacts again.
- The old credential sprawl trajectory quietly returns, and now is justified as operational necessity.

A drift-aware operating model catches this early, because the boundary signals change before anything becomes an incident:

- Override rate rises.
- Minted durations creep upward.
- Terminal collapse becomes unreliable.
- Off-road access pathways become routine.

The response isn't to "tell teams no." The response is: Treat it as a *platform defect.*

Either the curated actions are missing, the issuer needs an additional supported mode (still bounded), or the paved road needs integration work so that teams aren't forced off-road to do legitimate work.

The following table shows a practical way to frame drift signals.

Table 9.1: Four drift signal families and how to respond

Drift Signal Family	What You See	What It Usually Means	Likely Layer to Tune	Why It Matters
Pathway Pressure	Overrides, bypasses, clustered denied transitions	The paved road is incomplete, too slow, or missing a legitimate transition.	Curated actions, paved road, progression rules	If ignored, the teams create durable off-road routes.
Authority Pressure	Session duration creep, widening scope, terminal states not reached	The foundry is reproducing standing access or failing to collapse it cleanly.	Authority foundry, progression rules	Authority begins to persist faster than intent.
Interface Pressure	Raw control-plane usage returning,	The "steering wheel" is too weak or too loose.	Curated actions, capability catalog	Users reenter the engine room and ad

Drift Signal Family	What You See	What It Usually Means	Likely Layer to Tune	Why It Matters
	capability sprawl			hoc power returns.
Trust Pressure	Cross-boundary trust growth, emergency paths becoming routine	Trust is expanding faster than the structure is containing it.	Identity architecture, boundary design, paved-road coverage	Trust Inflation starts becoming the default operating condition.

The goal isn't to create another wall of telemetry. It's to identify which part of the constrained system is under repeated operational load.

What Drift Is Really Telling You

Drift isn't noncompliance. It isn't a morality tale about people taking shortcuts. It is the system telling you where reality is pressing against the current shape of the boundary.

Sometimes, the 'principal' is the problem. More often, the system is revealing one of four things:

- The paved road is incomplete.
- Authority doesn't collapse cleanly enough.
- The interface is still too raw.
- Trust is expanding through convenience faster than it's being structurally constrained.

It is why drift matters so much in high-speed environments. By the time the boundary actually breaks, the pattern has usually been visible for a while:

- In bypasses
- In longer sessions
- In raw control-plane fallback
- In trust edges that keep getting reused
- In "temporary" exceptions that stop feeling temporary.

Drift detection keeps you ahead of failure. But drift isn't the only teacher. Sometimes, the system teaches you through an incident.

Drift is the warning. An incident is the proof. A constraint patch is the learning mechanism.

Drift shows where the boundary is bending. An incident shows where the boundary has already failed, or where an unsafe path remained reachable long enough to create consequences. In both cases, the operating question is the same: which part of the decision space was too large, too permissive, too raw, or too slow to collapse?

Pathway Pressure asks whether the paved road is missing a legitimate path or allowing teams to skip a transition that must hold. Authority Pressure asks whether the foundry is minting power faster than the system can collapse it. Interface Pressure asks whether teams are falling back on raw control-plane capabilities because the curated action does not reflect the work. Trust Pressure asks whether cross-boundary authority is becoming normal again through roles, tokens, integrations, emergency paths, or inherited access.

The same taxonomy should govern both pre-incident drift and post-incident remediation. If Pathway Pressure appears before an incident, it indicates where a paved road may need to be widened or narrowed. If the same pressure appears in the incident review, it tells the organization what to patch. The incident did not create a new category of work. It made the existing pressure undeniable.

This changes the meaning of postmortem analysis. A serious incident should produce more than a timeline, a root cause, and a set of corrective actions. It should identify the pressure family that failed under load and the structural change required to prevent the same trajectory from remaining reachable. The organization should be able to say: this was not merely an outage, a bad grant, a missed cleanup step, or a one-off exception. It was evidence that a particular boundary did not hold under pressure.

The constraint patch is the response to that evidence. It is the change that makes the same unsafe trajectory harder, narrower, or unreachable next time. It may add a missing curated action, change how authority is minted, strengthen progression rules, force collapse on a terminal state, or replace a durable trust edge with a governed transition. The patch is complete when the system has been changed.

What Incidents Teach You: Constraint Patches as a Deliverable

In most organizations, incident response ends when the system is back online:

- Service is restored.
- Credentials are rotated.
- Dashboards are adjusted.
- A postmortem is published.

The team then moves on until the same class of failure returns under a different name.

Trajectory Governance Changes What "Done" Means

You still contain and restore. But if you stop there, you've only repaired the surface. The system's decision space—the set of futures it can reach—hasn't changed. The same unsafe trajectory remains reachable, waiting for the next outage, the next rushed deploy, and the next well-intentioned workaround.

In a trajectory-governed environment, every serious incident has a second deliverable: a constraint patch that deletes (or materially shrinks) the unsafe trajectory.

It isn't a reminder, a new rule, or a ticket to "be careful." It's a structural change to the following:

- The mechanism that turned intent into authority.
- The interface that expressed it.
- The progression rules that allowed it to continue.

It's why "architectural debugging" is a useful phrase here. The point isn't to debug people. It's to debug the system's allowed progression: the rails, the issuer, the curated actions, and the terminal states.

What a Constraint Patch Actually Looks Like

A good constraint patch has a particular feel. It doesn't read like "We tightened policy." It reads like "That move is no longer available." Sometimes, it means the following:

- That authority can't be minted the way it was minted before.
- The "fast path" everyone used under pressure is replaced by a paved-road action that actually works.
- A workflow now collapses by construction, rather than relying on someone to remember cleanup.

Appendix C provides a constraint root cause analysis (RCA) template and regression tests, designed to serve as a practical toolkit.

A practical way to use incidents is to convert the drift/pressure signals discussed in the previous section directly into patchwork. The drift signals introduced earlier (Table 9.1) — Pathway Pressure, Authority Pressure, Interface Pressure, and Trust Pressure — each call for a different kind of constraint patch. Drift tells you *where* the system is bending under reality. The constraint patch tells you *what to ship* so the system stops bending in the same place. The two sections share a taxonomy on purpose. If you can name the pressure family, you can name the patch. What follows is what each patch looks like in practice, in the same order as Table 9.1 introduced them

Patching Pathway Pressure

Pathway Pressure shows up as overrides, bypasses, and repeated denials on the same step. The team's instinct is to read this as a discipline problem. It isn't. Pathway Pressure usually means one of two things: either the paved road is missing a legitimate path, or teams are repeatedly trying to skip a gate that should hold. The first case needs capability. The second needs progression integrity. A bypass cluster is the most honest user feedback your platform will ever get. The paved road is losing because something about it doesn't fit the work; either it's too slow, it doesn't cover a real case, or it forces teams to drop down to the raw API to finish a job that the curated action almost handles.

The constraint patch follows the diagnosis. If the paved road is missing legitimate work, add capability: ship a curated action that completes the workflow end-to-end so teams don't have to leave the road. Reduce friction on the supported route before adding friction to the bypass. If the work has a real intermediate step that the transition model hasn't named, add that state and make the legitimate transition exist by construction. If the gate should hold, strengthen the progression rules so the skip becomes structurally unreachable rather than merely denied. Repeated denial is itself a pressure signal, and pressure that isn't relieved finds another route.

The hidden cost of leaving Pathway Pressure unpatched is what makes it urgent. Bypasses don't stay as bypasses. A reused role becomes a shared role. A one-off token becomes the token everyone knows works. A workaround that started as a one-time fix becomes the durable interface a downstream team builds against. By the time you notice, the bypass has become infrastructure, and the trust edge it created has become load-bearing. That's how an operational signal in week one becomes a structural problem by quarter three.

Patching Authority Pressure

Authority Pressure shows up as session-duration creep, scope widening, and terminal states that aren't reached reliably. All three are the same underlying failure expressed in different ways: collapse isn't holding. The system is good at minting authority and bad at ending it. Authority's half-life is the metric that matters here, and most organizations have never measured it.

The patches live at two layers. At the foundry, split coarse capabilities into narrower ones, enforce stricter maximum durations, and require renewals as explicit transitions rather than silent extensions. A timer can be extended; an explicit re-mint is a decision the system records. At the progression layer, make collapse structural rather than time-based: bind authority termination to workflow state, not to a wall clock. Force teardown on incident close, on artifact promotion, on task completion. Add "no silent persistence", every grant that should have collapsed and didn't is an edge in the decision graph the organization no longer governs.

Authority Pressure is the family most often misdiagnosed as a hygiene problem. It isn't. Lingering authority is the raw material trust inflation feeds on. Every "just leave it for now" is a future cross-boundary path that nobody designed.

Patching Interface Pressure

Interface Pressure shows up as raw control-plane usage rise inside what should be governed workflows, and as the capability catalog begins to sprawl with one-off actions that quietly become mini-admin powers. The signal is that the curated actions you shipped aren't covering the work.

The patch is to expand curated-action coverage and to discipline the catalog. Add intent-level parameters to existing actions so users don't have to drop to raw APIs to express what they actually want. Force reuse and refactoring under capability-design rules, no wildcard meta-capabilities, scope budgets on each new action, versioning so the catalog stays legible, and deprecation paths for actions that have grown beyond their original intent.

Interface Pressure and Pathway Pressure often appear together. The reason they're worth distinguishing in patch design is that the Pathway patch ships missing capability while the Interface patch refines existing capability.

Patching Trust Pressure

Trust Pressure shows up as cross-boundary trust growth — new cross-account roles, vendor tokens, inherited access paths — and as emergency pathways becoming routine. Both are signs that trust is expanding faster than the structure is containing it.

The patches here are the heaviest, because Trust Pressure is the layer where the other three families converge under sustained operation. At the identity-architecture layer, replace standing cross-account roles and long-lived vendor keys with bounded integration pathways: time-bounded, purpose-bound, workflow-scoped. Cross-boundary delegation becomes a governed transition rather than a durable shortcut. When emergency pathways become routine, treat that routine use as a product bug. Build a fast-mode transition into the model itself, a path that's faster than the normal path but bounded, observable, and collapsing. Add the curated actions that are missing for the high-pressure version of the workflow. Improve diagnostic pathways so engineers don't need standing privileged access to investigate. Then, and only then, tighten the conditions under which true emergency mode is permitted. The sequence matters: capability first, restriction second.

Institutionalized emergency is the final stage of trust inflation. It's where the organization has stopped pretending the boundary held and started treating the bypass as the boundary. A constraint patch here doesn't just shrink a trajectory; it restores the line between exception and norm. Without it, the rest of the model is a story the dashboard tells, not a property the system has.

What ties the four families together

Across all four, the patches share a shape. They don't add a reminder, a policy, or a ticket. They change what the system can express. A Pathway patch ships the missing capability. An Authority patch makes lingering grants impossible. An Interface patch refines the vocabulary teams use to act. A Trust patch makes cross-boundary delegation a governed transition rather than a durable shortcut.

Notice what's not in any of these patches: more enforcement on the surface that's already failing, more dashboards observing the same drift, more reminders to teams already working under pressure. The constraint patch always moves upstream to the issuer, the interface, the state machine, or the collapse rule. If the proposed patch merely tightens the same surface where the drift appeared, without changing the issuer, interface, or collapse rule, it is not a constraint patch. It is a slightly tighter version of the control that just lost.

Most organizations will patch constraints in a hybrid environment for years. Chapter 11 turns that into a transition playbook.

The New Definition of "Done"

In this model, an incident is closed only when …

- service is restored, and
- the pressure family has been named (Pathway Pressure, Authority Pressure, Interface Pressure, Trust Pressure, or some combination of them), and
- the unsafe trajectory has been structurally reduced or deleted. The patch must change the issuer, interface, collapse rule, or trust boundary that made the path reachable, and
- There is regression proof. The organization should be able to test that the deleted trajectory remains unreachable.

It turns security operations into a system that improves itself, not by promising perfect prevention, but by steadily shrinking the set of unsafe futures the environment can reach.

This is the new definition of done for security operations. Restore the business. Name the pressure. Patch the constraint. Prove the path is closed. Operations is done only when the same unsafe future is no longer reachable.

CHAPTER 10

The Sovereign CISO: Leading the Structural Shift

Once structural closure becomes the new definition of done, someone must own that loop institutionally. That is where the sovereign chief information security officer (CISO) enters. A sovereign CISO isn't a more powerful approver. The role is different, not bigger. A sovereign CISO is the executive steward of an organization's *control-plane constitution,* which is the set of structural rules that determine where authority can originate, how it can expand, and how it must collapse, especially under pressure and at machine speed.

Most organizations still ask the CISO to occupy one of two familiar positions. In one model, the CISO sits at the end of the delivery pipeline as the chief approver, signing off on changes, reviewing exceptions, and deciding whether the business can proceed. In the other model, the CISO sits downstream as the chief firefighter, detecting damage, coordinating response, and helping the organization recover to safety.

Neither role is sufficient for the environment this book has described. Trajectory governance demands a different security posture. The CISO is no longer only the organization's most senior gatekeeper or the executive sponsor of detection and response. Their job is to ensure that the organization can move fast without silently widening the set of unsafe futures it can reach. That means defining which decision categories are consequential. It means funding the platforms that make bounded authority easy to obtain. It means ensuring that unsafe trajectories are removed from the normal path of work. It means proving, under pressure, that speed is not being converted into trust inflation.

Risk doesn't only arrive as an attack. It accumulates as trust inflation through ordinary operational choices: broad roles that get reused because they work, cross-boundary integrations that become load-bearing, temporary access that never collapses, emergency shortcuts that persist after the emergency, and agents or automations that inherit more authority than their tasks require. In low-speed environments, people sometimes compensate for these weaknesses by hesitating. An engineer pauses. A manager asks for context. A security reviewer notices that something feels wrong. Human friction becomes an informal safety layer. But machine-speed systems don't provide that pause. This is why the CISO role has to change.

Chapter 8 described the engineering primitives that make trajectory governance real: transition models, authority foundries, curated actions, progression rules, and paved roads. Chapter 9 described how to operate those primitives under pressure by watching boundary pressure, detecting drift early, and turning incidents into constraint patches.

Here, I address what comes next: the *leadership layer* required to sustain that shift. Define what the sovereign CISO becomes accountable for, how the operating loop works, how progress is measured, and what must change inside the security organization when security is expected to produce structural outcomes rather than only policies, findings, and responses.
Producing structural outcomes requires security itself to change.

From Guardian of Infrastructure to Governor of Decision Authority

For most of the last two decades, the CISO mandate has been legible because the attack surface was legible. Security teams protected assets such as endpoints, servers, networks, APIs, and data stores. The program organized itself around domains that matched the system: AppSec for software, IAM for access; security operations center (SOC) for response; and governance, risk, and compliance (GRC) for assurance. The implicit promise was straightforward: Keep the bad things out, keep the important things safe, and respond when something breaks. That approach no longer describes or aligns with where risk is actually shaped.

In the world this book describes, risk isn't primarily a property of assets. It is a *property of authority:* who or what can decide, what actions are reachable as a result of those decisions, and how that authority expands or collapses across time. This is why the "sovereign" frame isn't rhetorical. It is a shift in what the organization is asking the CISO to govern.

A sovereign CISO doesn't stop caring about infrastructure. The shift is that infrastructure is no longer the center of the model. The center is the *decision authority*. Security leadership now has to govern the conditions under which systems, humans, workflows, suppliers, automations, and agents are allowed to take consequential action.

That shift becomes sharper as AI and agentic systems enter the enterprise. An agent is not merely another application to monitor. It is a decision system operating inside the authority surface that the organization has already created. It can call tools, retrieve data, update state, compose actions, and act at a speed and scale that makes human supervision a weak primary control. The security concern is no longer only whether the system is hardened. It is whether the system's authority is bounded.

From Asset Security to Decision Authority Security

Traditional security treats the system as an asset to be hardened: patched, segmented, monitored, and kept within an acceptable posture. Sovereign security treats the system as an actor whose authority must be bounded. It defines when the system may take consequential action, what conditions must be true before that action is valid, and which trajectories the system can never reach by design.

An asset-centric program tends to treat authority as a configuration detail: roles, permissions, policy documents, service accounts, and tokens. In a decision-centric security program, authority is the primary component of the attack surface; it's the currency that determines which system states and outcomes are even possible. If authority expands casually, the system can reach futures you never intended, even if every server is patched and every control is "green."

It is why autonomy fundamentally changes the leadership mandate. Autonomous systems don't merely "run faster software." They execute sequences of legitimate actions at a speed and scale that make human supervision irrelevant. If the CISO's governance model is still oriented around protecting assets and responding to events, the organization will discover that it's operating a system that can do far more than anyone can reliably oversee.

Decision authority security is the discipline of ensuring that the system's ability to decide and act is bounded by construction. It includes all decision authority—human, automated, and autonomous—because in modern environments those categories aren't cleanly separable. A developer pipeline is automation. A runbook trigger is automation. An LLM agent invoking tools is automation. The security question isn't: Is this actor human or AI? The question is, what authority can it hold, what can it do with it, and what sequence of moves can it compose?

From Static Risk to Trajectory Risk

Trajectory governance treats risk as motion. A sovereign CISO still cares about what exists now. But their program is accountable for something more demanding: what the environment can become under pressure, over time, through sequences of valid actions.

Modern failures often emerge through accumulation. Permissions grow one diff at a time. Trust relationships expand by reuse. Emergency access lingers and normalizes. Identity edges quietly dissolve boundaries the organization thought it still had. Vendor integrations become load-bearing. Agents inherit authority granted under assumptions that no longer apply.

A static state can show that a role is broad. It struggles to explain what that role has become: how it grew into a bridge, how many workflows depend on it, how many systems can exercise it, and how hard it is to reduce without breaking the business. That is trajectory risk and this is where trust inflation becomes a leadership concern.

Trust inflation happens when the supply of trust expands through broad roles, persistent tokens, inherited identities, unmanaged integrations, and cross-boundary relationships faster than the organization can constrain it. It is usually more rational than malicious. The sovereign shift is to treat trust inflation as a first-class risk signal. In a high-speed environment, trust inflation is what happens when every team optimizes locally and nobody governs what becomes reachable globally.

From Detection-First to Constraint-First

A classic CISO program invests heavily in detection and response. It isn't wrong. Detection and response remain essential because no environment is perfect. New attack paths appear. Controls miss things. People make mistakes. Vendors get compromised. Systems drift. The organization still needs to see, contain, restore, and learn.

The governance work moves upstream into constraints: the engineered conditions that determine which actions are expressible, which sequences are reachable, and how authority collapses when the mission ends. It's at this point where many security leaders feel discomfort, because it sounds like more control, more restrictions, and more friction.

But the constraint-first model isn't about adding gates. It is about removing unsafe moves from the system's design, so teams can move fast without accumulating unpayable risk.

A detection-first program sees unsafe movement and tries to react. A constraint-first program changes the environment so that known unsafe movement is harder, narrower, or impossible to express through the normal path of work. The difference isn't that detection disappears. The difference is that detection no longer carries the burden of governance on its own. The CISO stops trying to win a race against speed. They invest in making speed safe by default, instead.

The Executive Payoff of Trajectory Governance

The trajectory governance model is not only a better security model; it is also a better operating model for the business.

The first payoff is leverage.

In a traditional security program, security workload grows with business activity. More developers create more code paths. More cloud accounts create more configurations. More pipelines create more deployment events. More SaaS integrations create more access reviews. More agents create more tool calls, logs, approvals, and exceptions.

The result is a linear cost curve. Ten times the development activity creates ten times the findings, tickets, alerts, reviews, and analyst burden. This is why many CISOs feel permanently understaffed. The problem is not that the team lacks discipline. The problem is that the operating model scales with the number of things to inspect.

Structural governance changes that curve. When a paved road removes an unsafe trajectory from the normal path of work, the security team doesn't have to review, detect, triage, or remediate that failure mode every time another team encounters it. The platform does the work. The authority foundry mints bounded authority. The curated action narrows what can be expressed. The workflow enforces progression. Collapse removes authority when the mission ends.

A response-heavy model scales with incidents. A guardrail-heavy model scales with risky actions. A structural model like trajectory governance scales with the platform. Once the paved road works, every additional team that uses it inherits the governance outcome without requiring a proportional increase in security headcount.

The CISO stops trying to add people at the same rate the business adds systems, developers, integrations, and agents. Work moves into the path itself. The platform absorbs the governance pattern. The marginal cost of protecting the next team falls because the safe path has already been built.

The second payoff is predictable delivery.

In the old model, security approval often arrives late. A project moves forward, reaches a review point, and then discovers that access, deployment, data movement, vendor integration, or identity delegation requires exception handling. Security becomes the blocker, even when the objection is valid. In the sovereign model, the paved road is the pre-authorized business path.

If a project stays inside a governed pathway, the answer is already yes. The team doesn't need to rediscover policy, negotiate access, or wait for a manual exception. The boundary has been defined. The curated action exists. Authority is minted within scope. Collapse is built into the workflow. The approval has moved from the individual request into the design of the path. This changes the CISO's relationship with product and engineering. Security is no longer the function that delays delivery after intent has formed. It becomes the function that makes safe delivery predictable. The best security approval is the one the team never has to request because the approved path is already built into how work gets done.

This makes security an accelerant rather than a checkpoint. Product teams don't experience the CISO as the person who says no at the end. They experience security as the function that made the yes available earlier, faster, and with fewer surprises.

The third payoff is defensibility in the post-autonomy era.

Autonomous systems make perfect prediction impossible. A CISO will not be able to prove that every individual agent action was foreseeable, especially when agents reason probabilistically, compose tools, and act at machine speed. But the CISO can prove something more important: that the organization governed the trajectories available to those systems. The organization can show which decision categories were registered, which authority boundaries were enforceable, which curated actions existed, how authority was scoped, where collapse was required, and which unsafe trajectories were structurally unreachable.

Post-autonomy accountability will not only ask whether the organization detected a bad action. It will ask whether the organization gave autonomous systems broad authority without a bounded pathway. A sovereign program gives the CISO a stronger answer. The answer is: we governed what the system was capable of doing.

That is the executive case for trajectory governance. It lowers the marginal cost of safety, makes delivery more predictable, and provides leaders with a defensible architecture for systems whose individual actions can't be predicted in advance.

What a Sovereign Mandate Looks Like in Practice

A useful way to make this mandate concrete is to name the "consequential decisions" that define your organization's risk, because those decisions are where authority matters and where autonomy will apply pressure first. By consequential decisions, I mean the handful of decisions whose execution can materially change what the system can do next. This can be creating or expanding authority, crossing a trust boundary, moving sensitive data, causing an irreversible external effect, or changing the controls that govern the system itself.

Most organizations already have a handful of decision categories that dominate outcomes:

- Production change authority (e.g., deploy, rollback, config flips, and feature flags)
- Privileged access (e.g., break-glass, incident debugging, and administrative actions)
- Data movement (e.g., export, replication, training sets, and third-party sharing)
- Identity minting and delegation (e.g., issuing tokens, assuming roles, and cross-account trust)
- External side effects (e.g., sending messages, issuing refunds, changing entitlements, and disabling controls)

(See Appendix G for further information about decision categories.)

In an asset-centric model, these decision categories are scattered across teams, tools, and policies. In a sovereign model, these decision categories become the CISO's constitutional surface area: the set of decisions the organization must be able to govern structurally.

It doesn't mean that the CISO approves each decision. It means that the CISO is accountable for ensuring that the system enforces the boundaries around it by construction.

Most organizations are authorizing agents through delegated identity or by attaching them to human-scoped or legacy roles because they ship fast. That is a governance failure waiting to happen. If your program treats "AI security" as a separate workstream of model evaluations and prompt-injection defenses, you're missing the forest for the trees.

The failure isn't that the model is clever. The failure is that the model is attached to authority in ways that allow unsafe trajectories to become reachable.

In a sovereign model, a CISO's job is to define the *foundry rules* for these consequential decisions. When an agent or a developer asks to move data, the system shouldn't first look for an approval ticket. It should look for a paved road. If no paved road exists, the default shouldn't be "open a ticket." It should be that *this capability doesn't exist yet; ship it.*

The Three Non-Negotiables of Sovereignty

Sovereignty becomes real when you can answer three questions clearly, and when the system enforces your answers.

1. Where can authority originate?

In a non-sovereign environment, authority originates everywhere: Someone can write a new policy, mint a new token, attach a broad role to a workflow, or create a cross-boundary trust edge under pressure. This is the birthplace of trust inflation.

In a sovereign environment, certain classes of authority only originate through an explicit, constrained mechanism like an authority foundry that mints purpose-bound, time-bounded authority as part of a valid pathway. This is how you prevent "temporary" becoming permanent and "convenient" becoming load-bearing.

2. How can authority expand?

In a non-sovereign environment, expansion is incremental and invisible. A diff adds one permission. A new integration reuses an existing role. A workflow inherits a legacy trust edge. Expansion happens one reasonable step at a time, until the organization realizes it has erased the boundaries it thought it had.

In a sovereign environment, expansion is governed at the interface level. Teams don't expand raw authority by editing policies as the default workflow. They request capabilities and use curated actions. Expansion becomes visible because it requires extending the curated surface area, and not by quietly growing a policy document.

This is where trust inflation is either prevented or normalized. If expanding power is easier than requesting a bounded capability, trust inflation wins. If the paved road is faster and default, trust inflation loses momentum.

3. How does authority collapse?

This is the most underappreciated difference between a security program that "works on paper" and one that holds under speed.

Most organizations have an expiration. However, few experience collapse:

- **Expiration is passive:** Their access will end eventually, if the timer is correct and nothing extends it.
- **Collapse is structural:** A pathway ends because the workflow ends. Authority is dismantled because the mission is ending. A system returns to a safe terminal state as a property of execution, not as a cleanup task. An incident debugging role that expires after eight hours is better than a standing role. But if the incident ends after twenty

minutes, the remaining seven hours and forty minutes are still unnecessary authority.

A sovereign pathway binds collapse to the workflow state. When the incident closes, sessions terminate, tokens are revoked, and renewed access becomes a new explicit transition that is only valid if the incident remains active. Under autonomy and high-speed automation, collapse is what prevents your environment from becoming a museum of stale privilege and accumulated trust edges.

If authority does not collapse, it accumulates. If it accumulates, it becomes infrastructure. If it becomes infrastructure, the business will eventually depend on it. At that point, the organization is no longer making a clean security decision. It is trying to manage tech debt. That is why collapse is a leadership concern.

The Sovereign Operating Loop

A traditional security program often runs as a queue. Findings enter the queue. Exceptions enter the queue. Access requests enter the queue. Architecture reviews enter the queue. Incidents enter the queue. The security team works the queue as quickly and carefully as it can. This model creates effort. It doesn't always create closure. The same unsafe patterns return under different names. A broad role is reduced in one place and recreated in another. A risky deployment path is blocked in one pipeline and reappears in a second. A temporary access path is cleaned up after an audit and rebuilt during the next launch. A vendor integration is reviewed once and then quietly expands as more teams depend on it. The organization is busy, but the decision space has not changed enough.

The sovereign operating loop is designed to change this. It replaces ticket-driven remediation with a repeatable cycle of structural closure: identify which consequential decisions matter most, build the governed pathway, operate the boundary under pressure, and refine the system until the unsafe trajectory becomes harder, narrower, or impossible to reach. The loop is:

Define → Shape → Operate → Refine

Each step has a different purpose. Define the consequential decision. Shape builds the paved road. Operate watches to see whether the paved road holds under pressure. Refine ships the constraint patch when the system reveals that an unsafe trajectory is still reachable.

This loop is how a CISO turns the sovereign mandate from a leadership philosophy into institutional practice. It is also how security breaks the linear cost curve described in the previous section. The purpose of the loop is not to create more reviews, more dashboards, or more approval steps. The purpose is to move recurring risk out of human queues and into the design of the path itself.

Define: The Sovereign Contract

Define is where sovereignty becomes legible to engineering leaders and the broader organization. The goal is not to list every risk. It is to name the small set of consequential decisions the organization has committed to governing structurally.

The leader artifact for this step is the **Decision Authority Register**, or DAR. A DAR is not a risk register. A risk register tracks what the world might do to the organization (i.e., threats). A DAR tracks what the organization is allowing its systems to do (i.e., reachable trajectories). It names the decision categories where intent becomes authority: production change, privileged access, data movement, identity delegation, external side effects, supplier access, autonomous tool use, or any other action that can materially change what the system can do next. Each DAR entry should express the decision in operational terms:

Boundary → Curated Action → Collapse

The boundary defines the conditions under which the decision is valid. The curated action defines the supported way the work is done. Collapse defines how authority ends when the mission ends.

For example, "privileged production debugging" is not a vague risk category. It becomes a DAR entry when the organization can say: Production debugging authority may originate only inside an active incident workflow, for a specific service, for an approved purpose, through a supported diagnostic action, with automatic collapse when the incident closes.

DAR tells security what must be governed. It tells platform teams what must be built. It tells product teams what adoption must look like. It tells the board which high-impact decision categories are now governed by design rather than by hope, review, or after-the-fact cleanup.

The DAR also makes gaps visible. If a consequential decision is not in the DAR, the platform shouldn't mint authority for it by default. This doesn't mean the answer is no. It means the capability does not yet exist as a governed pathway. In most organizations, a missing pathway often becomes a ticket, an exception, or a legacy role. In a sovereign organization, a missing pathway becomes product work.

In the sovereign model, an exception is no longer treated as a request waiting for a senior approver. It is treated as evidence that a governed pathway is missing. Before the authority is made available, the organization must define the DAR row, boundary, curated action, and collapse condition that turn the request into a safe, repeatable path. This is how *Define* moves the organization away from ad hoc permission grants and toward trajectory governance. See Appendix F for more on operationalizing DAR.

This also gives applied security data science a clear target. The analyst's job is not to produce a generic risk dashboard. It is to help identify which decision categories are creating the most reachable consequence: where authority accumulates, where teams bypass the paved road, where trust inflation is rising, and where structural coverage is missing. Define tells the organization what must be governed.

Shape: Investing in the Paved Road

Shape is where policy becomes engineering. A conventional program may respond to a risky pattern by writing a standard, adding an approval step, buying a detector, or opening a recurring audit item. Those actions may help, but they don't necessarily change the supported path of work. *Shape* does.

In this phase, the CISO funds and sponsors paved roads that make governed execution faster than off-road execution. The paved road is not merely a safer option. It is the pre-authorized business path. If a team stays inside the governed pathway, the answer is already yes. They don't need to rediscover policy. They don't need to negotiate every access request. They don't need to wait for a manual exception. The boundary has been defined. The curated action exists. Authority is minted within scope. Collapse is built into the workflow. The approval has moved from the individual request into the design of the path.

Let's understand this with two examples.

The first is an **Authority Foundry**. Instead of granting standing production debugging roles, the platform mints short-lived, purpose-bound authority only inside a valid workflow. For example:

```
request_diagnostic_view(service, incident_id)
```

If the incident is not active, the authority doesn't exist. If the incident doesn't meet the required severity, the authority doesn't exist. If the engineer requests access outside the service scope, the authority doesn't exist. If the incident closes, the authority collapses. The team doesn't hold production debugging power. It moves through a pathway where the system creates bounded authority for a specific purpose and then dismantles it when the mission ends.

The second example is **Curated Action** deployment. Instead of allowing a broad identity to deploy however it wants, deployment becomes a supported action:

```
promote_to_production(artifact_digest)
```

The action is only valid when the artifact has a trusted provenance attestation, the service is in scope, the environment is approved, and the workflow is in the right state. This is not "detect bad deployments faster." It is "production deployment can't be expressed without the prerequisite state." This also shows the difference between a guardrail and a paved road. A guardrail catches an unsafe move after the system has allowed it to appear. A paved road changes the path so that the unsafe move is not a normal option.

Shape is where the CISO makes the most important budget decision in the sovereign model. The question is not only which tool to buy or which control to improve. The question is which paved roads deserve engineering capacity. Prioritization should be driven by evidence: DAR categories with the highest consequence, workflows generating the most exceptions, off-road paths that remain faster than the supported path, authority accumulating because a bounded alternative is missing, and autonomous workflows likely to scale an unsafe pattern.

This is where security, platform, product, and applied security data science have to work together. Security defines the boundary. Platform builds the mechanism. Product makes the path usable. Data science shows where the decision space is bending and where investment will have the most leverage. *Shape* is not successful when the paved road merely exists.

Operate: Reading Boundary Pressure

Operate is how the organization learns whether the paved road is holding under real conditions. A sovereign system is not operated only by watching logs for attackers. It is operated by watching whether the governed path remains the easiest path when pressure rises. Pressure is where governance is tested. A launch deadline approaches. An incident escalates. A vendor integration becomes urgent. A data export is needed for a business review. A team wants to connect an agent to an existing role because creating a new bounded path will take longer. These are the moments when trust inflation returns. Not because people are careless. Because the system offers an easier unsafe path than the governed one.

Boundary pressure is the early warning signal. Boundary pressure shows up as repeated exceptions, manual overrides, legacy roles that refuse to die, emergency access that becomes routine, teams bypassing curated actions, or agents being attached to broad identities because the task-scoped pathway doesn't yet exist. In the sovereign model, boundary pressure triggers product behavior, not policing. Security asks include:

- Why is the paved road losing? Missing curated action? Too much latency? Wrong boundary assumption?
- What new constrained pathway needs to be shipped, so that the exception stops being rational?

This is why operating the boundary feels less like compliance enforcement and more like running a platform where you treat bypasses as user experience (UX) signals, and not just a list of security exceptions.

Refine: Constraint Patches and Time-to-Closure

Refine is where sovereignty compounds and risk is actually deleted.

In a vigilance-based program, a postmortem often ends with "more training" or "better alerts." In a sovereign program, an incident is evidence that an unsafe trajectory was reachable. The second deliverable is non-negotiable: a constraint patch. A *constraint patch* is a structural change to the boundary, a curated action, foundry rules, or collapse semantics that makes that class of failure harder or impossible to reach again.

Consider break-glass access. The incident is not only that someone retained access too long after an emergency. The deeper issue is that the system allowed emergency authority to outlive the emergency. A weak remediation says: "Remind responders to revoke access when the incident ends." A stronger guardrail says: "Alert if access remains after the incident closes." A constraint patch says:

"Bind authority collapse to the incident state. When the incident closes, sessions terminate, tokens are revoked, and renewed access becomes a new explicit transition that is only valid while the incident remains active, with a hard time-to-live backstop."

The unsafe trajectory has not merely been documented or detected. It has been narrowed or removed. Refine also gives the CISO one of the most important operating metrics in the sovereign model:

Mean Time to Structural Closure.

Mean Time to Structural Closure measures the time from discovering a reachable unsafe path to shipping the constraint patch that deletes or materially bounds that path. This is not the same as mean time to remediate. A vulnerability may be patched. A ticket may be closed. An incident may be restored. But if the same class of unsafe trajectory remains reachable, the organization has not structurally improved.

It is the metric that turns incidents, exceptions, and bypasses into compounding improvement. *Refine* should also track constraint patches shipped per period, collapse failures corrected, off-road paths deprecated, and DAR categories moved from policy coverage to structural coverage. These are not activity metrics. They are evidence that the decision space is changing.

A sovereign program doesn't mature by becoming better at explaining why the same unsafe path keeps appearing. It matures by making that path disappear.

The Scoreboard: What the Sovereign CISO Measures and Reports

Traditional security reporting is built around activity and condition: vulnerability counts, control coverage, compliance status, mean time to detect, mean time to respond, patch aging, and audit exceptions. These measures are not useless. They convey something about the organization's posture and operational discipline.

Trajectory governance is not primarily an effort to find more problems. It is an effort to make certain classes of failure unreachable by construction, and to keep that claim true as the organization moves fast, delegates authority, connects systems, and accumulates trust. The scoreboard, therefore, shows which consequential decisions are governed by design, whether bounded authority is faster than unsafe workarounds, whether trust inflation is shrinking or growing, and whether the security program is breaking the linear relationship between business growth and security workload.

The scoreboard has two layers. The first layer is for the board, CEO, and executive peers. It reports structural outcomes that the business can rely on. The second layer is for the operating system. It gives security, platform, product, architecture, and operations teams the signals they need to steer the next investment. The executive layer shows whether the business is getting safer as it moves faster. The operator layer shows where the model is bending under pressure. A sovereign CISO needs the first to fund and explain the shift. They need the second to keep the shift real.

What the Executives Should See: Structural Outcomes, Not Compliance Theater

Executives don't need a tour of the security toolchain. They need to understand whether the organization can move at the speed the business requires without silently expanding its blast radius. That means the sovereign CISO reports in terms of decision authority governed, unsafe trajectories removed, speed gained through paved roads, and trust inflation reduced.

The executive scoreboard should be small enough to remember and strong enough to steer investment. Four metrics matter most.

1. **Decision Authority Coverage:** Decision Authority Coverage measures how much of the organization's consequential decision surface is governed structurally. It is the share of Decision Authority Register categories that have moved from policy, review, and snapshot-based governance into enforceable boundaries, curated actions, and reliable collapse.

   The question it answers: What fraction of our most consequential levers are governed by trajectory governance today?

A consequential decision on the DAR counts as structurally governed, only when the category is "complete" with the following conditions, not when it is merely documented (definition of done):

- *Boundary* is enforceable in the control plane (the unsafe move is *unreachable,* not just "disallowed by policy").
- A *curated action* exists and is the normal interface for doing the work. (Teams don't have to go off-road to get it done).
- *Collapse* is automatic and reliable. (It has terminal-state integrity: Authority actually collapses when the mission ends.)
- As the program matures, the paved road actually handles the great majority of executions for that category.

Coverage should eventually reflect operational reality, rather than design intent. A category isn't truly "covered," because the path exists. It is covered when a certain percentage of teams (i.e., the threshold) actually use it. Defining a threshold turns DAR coverage from a design claim into a governance claim. It forces the program to earn coverage through usability, and not paperwork.

An executive update might sound like this:

Production deployment and break-glass access moved to structural coverage this quarter. Data export is partially covered, but adoption remains below threshold because two business units still rely on legacy export workflows. Identity delegation remains policy-covered but not structurally covered, and it is now the next DAR priority.

2. **Mean Time to Authority:** Mean Time to Authority measures how quickly teams can obtain bounded authority through the paved road. This is the sovereignty speed dividend. It proves whether the safe path is fast enough to win under real delivery pressure.

   The question it answers: Are we making the safe path the fastest path?

   If the governed path is slower than the unsafe path, teams will route around it. They may use a standing role, reopen an old exception, reuse a legacy service account, or ask for broad access because the precise path takes too long. That is not only a usability problem. It is how trust inflation returns.

   Mean Time to Authority should be reported the way operators experience it: median and P90 (90th percentile). The goal is not merely to show that bounded authority exists. The goal is to show that bounded authority is faster than the legacy alternative.

An executive update might sound like this:

Foundry-issued production debugging authority now arrives in under two minutes at P90. The legacy ticket path remains measured in hours. Adoption rose because the bounded path became faster than the workaround.

3. **Trust Inflation Trend:** Trust Inflation Trend measures whether the organization's supply of broad, persistent, reusable trust is growing or shrinking as the business scales. This includes standing administrative roles, long-lived tokens, unmanaged trust edges, broad service accounts, inherited identities, vendor integrations, cross-account relationships, and agent permissions that persist beyond their task.

   The question it answers: Is our cumulative risk debt shrinking or growing as we scale?

   The metric should be segmented by category:

   - human authority,
   - workload authority,
   - agent authority,
   - supplier or vendor authority,
   - cross-boundary authority.

   This segmentation highlights whether trust inflation hides in one area while another appears to improve. A company may reduce human standing access while increasing workload trust. It may tighten cloud roles while expanding SaaS integrations. It may improve break-glass collapse while attaching agents to broad legacy identities. A single blended score would hide the drift.

   Trust Inflation Trend shows whether the sovereign program is replacing cheap trust with purpose-bound, collapsing authority.

   A useful executive update might sound like this:

   Overall blast radius contracted this quarter as standing administrative roles were replaced with foundry-minted authority. Trust inflation is still rising in supplier integrations and agent delegation. Those two categories are being added to the next DAR cycle.

4. **Security Leverage Ratio:** Security Leverage Ratio measures how many governed executions occur through paved roads for every manual security intervention required. It is the metric that shows whether security is breaking the linear cost curve.

   The question it answers: Is security workload decoupling from business growth?

   In a traditional program, more business activity creates more security work. More developers create more reviews. More pipelines create more findings. More cloud accounts create more posture drift. More agents create more logs, approvals, and exceptions. Security workload rises with the number of things to inspect. A sovereign program should bend that curve.

   As paved-road adoption increases, more work should move through governed pathways without requiring proportional increases in tickets, reviews, escalations, and analyst intervention. The security team still works. But its work shifts from repeated inspection to platform improvement.

   For example, production deployment might show that 8,000 governed deployments occurred through the curated path while only 40 required manual security involvement. Break-glass access might show 300 foundry-issued sessions with 12 manual exceptions. Agent tool use might show the opposite: high manual review volume because the governed pathway is immature.

   The ratio is not meant to eliminate human review. Novel, ambiguous, or high-consequence decisions may still require explicit review. The purpose is to show whether repeatable work is moving out of human queues and into governed pathways. A rising Security Leverage Ratio means the security program is scaling through design. A flat or declining ratio means the business is growing faster than the control model.

Together, these four executive metrics show whether the sovereign model is producing its promised business outcomes. Decision Authority Coverage shows which consequential levers are governed by design. Mean Time to Authority shows whether the safe path is fast enough to win. Trust Inflation Trend shows whether growth is expanding or contracting blast radius. Security Leverage Ratio shows whether security workload is decoupling from business scale. This is the executive scoreboard. It shows whether the organization is making unsafe futures less reachable while making safe delivery faster.

What Operators Need to Steer the Machine

Internally, the goal is to steer. Security, platform, product, and operations teams need to know where paved roads are losing, where collapse is failing, where trust is inflating, where reachability is expanding, and where a constraint patch should be funded next. Vulnerabilities, misconfigurations, drift alerts, policy failures, and incidents become telemetry. They show where authority is leaking, where trust is accumulating, where teams are bypassing the supported path, and where the system is still able to express unsafe trajectories. No single tool produces this view. Vulnerability scanners show findings. Identity tools show entitlements. Cloud platforms show resources and policies. CI/CD systems show workflows. Ticketing systems show approvals and exceptions. SOC tools show events. Each view is useful. None of them alone explains how authority moves.

The operating scoreboard depends on connecting those sources into decision-space intelligence.

The following are the four operator-level metrics for the operating loop:

1. **Boundary Pressure (Bypass and Override Rate):** Boundary Pressure measures where teams are bypassing, overriding, or routing around the paved roads. It tracks off-road executions: legacy roles, manual overrides, exception tickets, break-glass overuse, direct policy edits, unmanaged integrations, and agent workflows attached to broad identities instead of task-scoped authority.

   The question this answers: Where is the business pushing against our boundaries because the paved road is missing, slow, or too rigid?

   The crucial point is that bypass isn't treated here as a disciplinary failure. It is treated as product feedback. A bypass means the paved road is losing to reality. It may be too slow, too narrow, too fragile, or missing a real operational need. The boundary may be modeled incorrectly. The curated action may not exist. The legacy path may remain easier because teams already know how to use it. Boundary Pressure turns bypass into engineering work.

   It tells product teams where adoption is failing. It tells platform teams where the authority foundry needs a new minting pathway. It tells security leaders where trust inflation is reentering through convenience. A useful

Boundary Pressure report doesn't only count exceptions. It classifies why they occurred:

- missing curated action,
- slow authority minting,
- incorrect scope boundary,
- unavailable workflow state,
- unreliable collapse,
- legacy path still easier,
- agent or automation use case not yet supported.

This prevents the organization from falling into the "more training" reflex. The goal is not to remind teams harder. The goal is to make the governed path win.

2. **Terminal-State Integrity (Collapse Success):** Terminal-State Integrity measures whether authority actually collapses when the mission ends. This metric captures one of the most important differences between expiration and collapse. Expiration waits for time to pass. Collapse binds authority to workflow state.

   The question this answers: Does authority reliably vanish after use, or are we accumulating stale trust edges?

   Break-glass access should terminate when the incident closes. Production debugging authority should disappear when the diagnostic session ends. Vendor access should end when the integration scope expires. Agent tool authority should collapse when the task reaches its terminal state. Terminal-State Integrity shows how often that actually happens.

   It tracks the percentage of authority grants that collapse correctly, the number that linger, the time between mission completion and authority termination, and the reasons that collapse failed. Collapse failures are not minor hygiene issues. They are the mechanism by which temporary authority becomes standing authority.

3. **Mean Time to Structural Closure (Constraint Patch Throughput):** Mean Time to Structural Closure measures how long it takes the organization to move from discovering a reachable unsafe path to shipping the constraint patch that deletes or materially bounds that path. This is the *Refine* metric. It is not the same as mean time to remediate.

The question this answers: Are we learning structurally, or are we just producing remediation churn?

Mean time to remediate often measures whether a ticket was closed, a vulnerability was patched, a configuration was corrected, or an incident was restored. However, these remediation leave the underlying trajectory intact. Mean Time to Structural Closure measures whether the system improved. A constraint patch may add a curated action, tighten a foundry rule, add a workflow state, remove an off-road path, or bind collapse to a terminal condition.

The metric should be paired with constraint patches shipped per period. Mean Time to Structural Closure shows learning speed. Constraint patches shipped shows learning throughput. Together, they show whether the organization is converting incidents, near misses, exceptions, and bypasses into structural improvement.

4. **Reachability Ratio (Maximum Reachable Damage):** Reachability Ratio measures the maximum surface area an entity can affect if its authority is misused. It is a practical way to quantify blast radius without relying only on subjective severity labels. An entity may be a human identity, workload identity, service account, CI/CD runner, SaaS integration, vendor token, or agent. Relevant reachability may span identity, network, data, resource, deployment, and control-plane relationships.

   This is a graph-native calculation of the "Blast Radius" available to an entity across Identity, Network, and Resource planes. It is the ratio of Reachable Assets () to Total Assets (). More mature versions weigh reachability by sensitivity, authority type, environment, and consequence. A lower Reachability Ratio means a narrower, more bounded future. This is one of the most practical ways to quantify whether paved roads are actually reducing blast radius without relying on subjective risk ratings.

   The question this answers: What is the maximum surface area of the organization this entity can impact if its authority is misused?

   If a CI/CD runner can reach every production account, its reachability ratio is too high. If a vendor integration can access broad datasets unrelated to its purpose, its reachability ratio is too high. If an agent inherits a role that touches identity, data, and external communication, its reachability ratio is too high. A sovereign program should drive reachability down for high-impact identities and workflows.

Reachability Ratio helps prove that the environment is not merely cleaner on a dashboard. It is less capable of producing broad damage if something goes wrong. (See Appendix H for information about the Reachability Ratio Toolkit.)

Bringing It Together

A sovereign scoreboard works best when it's reported as a short "three-part truth" that keeps the narrative focused on the following outcomes:

1. **What we made impossible:** Which classes of failure were eliminated this quarter? Which DAR categories moved to true structural completion?
2. **Where pressure is building:** Where are bypasses, collapse failures, trust expansion, or excessive reachability telling us the current model is losing to reality?
3. **The speed dividend we created:** How much faster can teams now move because the paved road is the shortest path?

This format keeps the metrics from becoming a reporting exercise. The numbers serve the governance narrative. The governance narrative drives engineering investment. The engineering investment changes what the system can express. Decision Authority Coverage improves when boundaries become enforceable, curated actions become normal interfaces, collapse becomes reliable, and adoption crosses the threshold. Mean Time to Authority improves when bounded authority becomes faster than tickets and workarounds.

Trust Inflation Trend improves when standing power is replaced by purpose-bound, collapsing authority. Security Leverage Ratio improves when more business activity flows through governed pathways without proportional growth in manual security review. Boundary Pressure improves when bypasses become product backlog, and the paved road starts winning. Terminal-State Integrity improves when authority disappears with the mission instead of lingering as debt.

Mean Time to Structural Closure improves when the organization ships constraint patches quickly after discovering reachable unsafe paths. Reachability Ratio improves when identities, workloads, vendors, and agents can affect less of the environment by default.

That is the scoreboard of a sovereign CISO. It is not a posture dashboard.

It shows whether the organization is eliminating classes of failure, whether safe paths are fast enough to win, whether trust inflation is being reduced, whether autonomy is being bounded, and whether the security function is producing leverage instead of simply absorbing scale.

A scoreboard like this is only credible if the organization can move the numbers. They do not move through reporting discipline alone. They move when the security function changes the system: funding paved roads, tightening foundry rules, improving curated actions, repairing collapse, deprecating off-road paths, and shipping constraint patches.

Once security is expected to report structural outcomes, it needs an organization capable of producing them. The next section turns to that institution: the roles, skills, ownership model, and operating cadence required when security must ship governable systems rather than only policies.

Building the Institution: Organization, Roles, and the New Security Work

To move from posture to constraint design, the security organization itself must be reengineered. You can't achieve structural outcomes with an organizational chart designed for manual vigilance. If your team is optimized to find and "nag", it will remain trapped in the posture cycle even if everyone intellectually agrees with everything in this book.

Security moves from being a reviewer of work to a *producer of constraints, paved roads, decision-space intelligence, and structural closure*. Now, it doesn't mean that reviews or detection work disappear. It means that reviews stop being the primary control. They become telemetry: Signals that tell you where the paved road is missing, where authority is being minted off-road, and where trust inflation is accumulating faster than you can constrain it.

The Definition-of-Done Pivot

In a posture-based organization, security work is often defined by task completion: Run the scan, assign the findings, verify the patch, document the exception, and close the finding. In a sovereign model, those activities still happen, but their purpose changes. They stop being the end goal and become *inputs into engineering.*

A practical way to institutionalize the shift is to change the definition of done across major security functions. The old definition of done is often evidence that security work was performed. The new definition of done is evidence that unsafe reachability was reduced. Activities become telemetry; engineering becomes the control, as shown in the following table:

Table 10.1: Security functions, posture output vs. sovereign output

Function	Posture/Vigilance (Old Output)	Sovereign/Engineering (New Output)
Vulnerability management	Patch a specific common vulnerabilities and exposures (CVE) to "green" on a dashboard.	Remediate CVEs prioritized by reachability to consequential authority and use the aggregated feed to surface which asset classes recur often enough near authority to warrant a constraint patch by platform or architecture.
Identity and access	Perform manual reviews and periodic attestations. ("Remove stale grants.")	Engineer collapse authority pathways so standing access does not accumulate by default. Replace broad, persistent grants with purpose-bound, time-bound, workflow-bound authority.
Network/segmentation	Manage firewall rules based on "who needs what?"	Drive reachability ratio down by design. Constrain cross-boundary movement so trust inflation cannot accumulate cheaply through network proximity, inherited trust, or broad service identity. Connectivity requires an explicit, reviewable declaration rather than a ticket-driven exception

Function	Posture/Vigilance (Old Output)	Sovereign/Engineering (New Output)
AppSec	Run static application security testing (SAST)/dynamic application security testing (DAST) and hand findings to development teams.	Design secure-by-default patterns, libraries, and abstractions that platform teams adopt as paved-road primitives. Move from finding-by-finding SAST/DAST triage to making safe patterns the default in normal development workflows, so unsafe code paths become harder to express in the first place.
Cloud security posture management	Detect misconfigurations and open tickets for teams to remediate.	Treat configuration as boundary integrity, not as compliance. Detect drift as a trajectory signal; which boundaries are bending, which supported paths are being bypassed, which configurations recur often enough to indicate what paved road doesn't cover. Hand the pattern signal to platform engineering as input to the next constraint, default, or module that removes the unsafe variant.
Security operations	Detect, triage, contain, and restore after incidents.	Operate the boundary, not just the alert queue. Continue to detect and respond, but track bypasses, collapse failures, drift, and unsafe trajectories that remain reachable. Treat every meaningful incident as producing two deliverables: restoration for the business, and the structural signal that feeds the constraint patch.

Function	Posture/Vigilance (Old Output)	Sovereign/Engineering (New Output)
AI and agent security	Evaluate models, test prompts, monitor outputs, and review agent behavior after deployment.	Govern authority, not just behavior. Continue evaluating models and monitoring outputs, but specify the decision boundary each agent is sanctioned to operate within: task scope, tool scope, workflow state, and terminal conditions. Partner with platform engineering to make those bounds enforceable by construction.

Security does not stop scanning, reviewing, detecting, or reporting. It stops treating those activities as the final product.

The product becomes …

- paved roads,
- authority foundries,
- constrained interfaces,
- collapse semantics,
- and constraint patches that permanently narrow unsafe reachability.

Table 10.1 tells half the story. It names how each security function changes what it ships, but every row in the table now ends with the same implicit handoff: the security function produces signal and design intent, and someone else ships the structural change. That someone else is not a single team. It is a set of roles, some new, some evolved from existing functions, that together carry the sovereign work the seven functions above can no longer do alone. The rest of this section names those roles, what each one owns, and how they fit together.

Product Management Is Not Optional

A paved road is not adopted because it exists. It is adopted because it works under pressure. This is also where many security platform efforts fail. A team publishes a secure module, documents a new workflow, announces the paved road, and declares victory. The road technically exists, but teams continue to route around it because the old path is faster, clearer, more flexible, or better understood.

Adoption is not an engineering property. It is a product property. Trajectory governance only works when the paved road wins during a deadline, during an incident, during a launch, during a migration, and during the messy operational moments when teams are most tempted to widen authority. That requires product ownership.

The paved road needs a roadmap. It needs an adoption strategy. It needs usability goals. It needs onboarding. It needs support. It needs migration paths. It needs a deprecation plan for off-road routes. It needs metrics that show whether teams are actually using it. It needs a product owner who owns mean time to authority metric. This is why product management belongs inside the sovereign institution. A security team may define the boundary. A platform team may build the mechanism. But product management makes the governed path usable enough to become the default. Without product ownership, paved roads become shelfware. With product ownership, they become pre-authorized business paths.

A sovereign institution needs product managers (or product-minded owners) for governance of the pathways. They need PMs for paved roads as a product. (See Appendix I for more information on a security PM's manifesto.)

Applied Security Data Science: Making Decision Space Legible

A sovereign institution also needs an analytical capability that many security organizations do not yet have. The CISO cannot govern decision authority with disconnected dashboards. A vulnerability dashboard shows findings. An identity tool shows entitlements. A cloud platform shows resources and policies. A CI/CD system shows workflows. A ticketing system shows approvals and exceptions. A SOC platform shows events. Each view is useful. None of them, by itself, explains how authority moves through the organization.

Applied security data science turns fragmented operational data into decision-space intelligence. It connects vulnerability data, identity data, cloud permissions, CI/CD records, SaaS integrations, ticketing history, incident data, exception patterns, access reviews, and workflow activity into a practical view of reachability, authority accumulation, trust inflation, boundary pressure, and structural closure.

This capability is not about predicting every attack or producing a perfect graph of the enterprise. Applied security data science shows where unsafe trajectories are becoming reachable, where authority is accumulating faster than responsibility, where paved roads are losing to off-road paths, where trust inflation is rising, and where structural investment should go next.

It also changes how security prioritizes. A traditional reporting function may show the number of critical vulnerabilities. Applied security data science identifies which vulnerable systems also sit near privileged authority. A traditional identity report may show the number of privileged accounts. Applied security data science identifies which identities are accumulating permissions over time and which ones create cross-boundary reachability. A traditional ticketing report may show exception volume. Applied security data science shows which exceptions reveal missing paved roads.

This function is where the security organization learns to see its decision space. Without it, the sovereign program risks becoming another governance ceremony. Leaders may define DAR categories. Product teams may build paved roads. D-SOC may monitor boundary pressure. But the organization will still struggle to know where the next structural investment should go.

Applied security data science provides the CISO with evidence for funding decisions. It gives product managers adoption and bypass signals. It gives platform teams the patterns that need new curated actions. It gives decision security architects a map of where boundaries are bending. It gives D-SOC a way to distinguish isolated alerts from signals of structural drift.

In a sovereign security organization, analytics is how the institution sees its decision space.

Platform Engineering and Decision Security Architecture: Who Builds Sovereignty

Security can define what must be unreachable. But most of what makes that true lives in the platform layer. The platform defines how authority is minted, what curated actions exist, how sequences are enforced, how collapse happens, how developer workflows are shaped, and how paved roads become the default path of work. This means sovereignty cannot be the program of security alone. It must be co-owned with platform engineering, product management, enterprise architecture, operations, and the teams that own the workflows where consequential authority is exercised.

A clean division of accountability looks like this:

Table 10.2: Roles in the sovereign security organization

Capability	Owns	Practical Output
Security leadership	The constitution	DAR boundaries, collapse requirements, board narrative, funding priorities, and the definition of which decision categories require structural governance.
Decision security architecture	Translation	Authority foundry rules, curated action patterns, progression requirements, boundary models, collapse semantics, and control-plane design patterns.
Applied security data science	Decision-space intelligence	Reachability models, trust inflation analysis, boundary-pressure signals, paved-road adoption evidence, DAR prioritization, and structural-closure evidence.
Platform engineering	Machinery	Foundry implementation, curated actions, progression logic, control-plane enforcement, paved-road reliability, and integration into developer and operator workflows.
Product management	Adoption	Paved-road roadmap, usability, mean time to authority, migration planning, adoption metrics, and deprecation of off-road paths.
D-SOC	Feedback loop	Boundary pressure, drift review, collapse failures, trust inflation signals, incident-to-constraint-patch conversion, and structural learning from operational events.
Enterprise architecture	Coherence	Reusable boundary patterns across business units, vendors, platforms, identity systems, data flows, and shared services.

No single function can produce sovereignty alone.

Security leadership can declare a boundary, but platform engineering must make it enforceable. Platform engineering can build a workflow, but product management must make it usable. Product management can drive adoption, but applied security data science must show where adoption is failing and where trust is accumulating. D-SOC can operate the boundary, but decision security architecture must define what the boundary is supposed to be. Enterprise architecture can standardize patterns, but it must do so without becoming a slow approval body that teams route around.

The decision security architect and D-SOC are especially important because both are often missing in traditional security organizations. The decision security architect works upstream. This role translates constitutional intent into control-plane design. It defines the authority foundry rules, curated action surfaces, workflow progression requirements, and collapse semantics that make consequential decisions governable.

D-SOC works in motion. It watches whether the designed boundary holds under operational pressure. It monitors bypasses, collapse failures, drift, trust inflation, reachability expansion, and incidents that reveal a missing constraint patch. One defines the intended boundary. The other operates the boundary under pressure.

Applied security data science supports both. It provides the evidence that shows which boundaries matter most, which ones are bending, and which structural investments will reduce the reachable consequences. Together, these roles create the institutional machinery of sovereignty.

The operating rule is simple: when the business needs a high-impact capability, and no paved road exists, the default is not to route the request to a senior approver. The capability becomes product and platform work. A new high-impact capability becomes available only when the DAR row is defined, the boundary is clear, the curated action exists, and collapse is built into the workflow.

Where Enterprise Architecture Fits Without Becoming a New Gate

Most enterprises already have an EA function. In a sovereign model, EA is valuable precisely because trajectory governance is performed cross-domain: It is performed across the domains of identity, data movement, production change, delegation, and vendor trust; none of them are governed by merely one team.

EA fits best as a coherence engine:

- Standardizing enterprise boundary patterns (standardizing what "production deploy," "break-glass," "export," "delegation" mean)
- Ensuring curated actions align with reference architectures and shared platform primitives
- Preventing "local paved roads" that work in one business unit (BU) but inflate trust across the enterprise
- Aligning vendor/platform strategy to constraints that are enforceable and not merely documented

If enterprise architecture becomes a slow approval body, it becomes an off-road generator. Teams under pressure will route around it. Trust inflation will accelerate through local workarounds..

EA should standardize constraints that the platform can enforce automatically. It should not become the reviewer of every individual change. EA success is measured by the reusability of enforceable patterns, not by the number of reviews performed. Enterprise architecture defines the reusable constraint patterns. Platform implements them. Security governs them. Product drives adoption. D-SOC operates the pressure signals. Applied security data science shows where the patterns are working and where they are failing.

SOC → D-SOC: Operating the Boundary, Not Just the Alert Queue

D-SOC does not replace the SOC. It extends the SOC's mission from detecting attacks to operating decision boundaries. A trajectory-governed environment still needs detection and response; however, the operational object changes. D-SOC focuses on:

- Boundary pressure (of bypasses, overrides, and exception volume)
- Drift signals (where authority and trust edges are expanding)
- Terminal-state integrity (collapse success vs. lingering authority)
- Trust inflation trend (where "cheap trust" is reaccumulating)

Most importantly, the definition of "done" also changes: Restoration is necessary. Structural closure is mandatory. Every meaningful incident yields two deliverables:

1. Restoration for the business

2. A constraint patch that deletes or sharply bounds the reachable unsafe trajectory.

This operational definition is what makes sovereignty compound rather than churn.

What Changes on Monday

A leader does not need to rebuild the entire security organization at once. The first moves should be concrete enough to begin immediately and consequential enough to change how the institution operates.

The starting point is not a new dashboard, a new committee, or a larger exception queue.

The starting point is making one slice of decision space visible.

1. **Make one decision space visible.**

   Pick one consequential decision category where the organization already feels pressure: production deployment, break-glass access, data export, identity delegation, vendor access, or agent tool use. Map how the work actually happens today. Identify who or what can initiate the action, where authority originates, how it expands, which systems and data become reachable, which shortcuts teams use, where exceptions accumulate, and where authority fails to collapse. The output is not a generic risk report. It is a decision-space map: a practical view of what becomes possible from this workflow.

2. **Stand up decision-space analytics.**

   Assign an applied security data scientist, or a small applied analytics capability, to the first decision category. The work is to connect the relevant identity, cloud, CI/CD, vulnerability, ticketing, incident, SaaS, and workflow data into one usable view. This capability should show where authority originates, where it widens, where teams bypass the supported path, where trust inflation is accumulating, and which unsafe trajectories remain reachable. This is how the CISO stops steering from disconnected dashboards and starts steering from decision-space intelligence.

3. **Create the first DAR row from the map.**

   Once the decision space is visible, convert the highest-consequence workflow into a Decision Authority Register entry. Define the boundary, the curated action, and the collapse condition. The DAR row should not be abstract. It should describe the specific authority the system is allowed to mint, the conditions under which that authority is valid, and how it ends when the mission ends.

4. **Create a joint structural-governance backlog.**
   Security, platform, product, applied security data science, and operations should share one backlog for structural security work. The backlog should be fed by the decision-space map, D-SOC boundary pressure, drift reviews, collapse failures, trust inflation signals, high-volume exceptions, and off-road paths that teams still use under pressure. This keeps the backlog grounded in how the environment actually operates, not in abstract control aspirations.
5. **Name product ownership for the paved road.**
   The paved road needs a product owner, not just an engineering owner. That person owns adoption, usability, mean time to authority, migration planning, and deprecation of off-road paths. Their job is to understand why the unsafe path is still faster, clearer, or more reliable than the governed path, and to make the governed path win under real delivery pressure.
6. **Make one path pre-authorized.**
   Choose one high-volume workflow and define what it means for the path to be pre-authorized. If a team uses the curated action, stays inside the boundary, and accepts automatic collapse, no separate security approval should be needed. This proves that sovereignty is not a slower approval model. It is a faster delivery model.
7. **Make co-ownership explicit.**
   Security defines the DAR boundary and governance requirements. Platform engineering ships the mechanism. Product owns adoption. Applied security data science supplies decision-space intelligence. D-SOC operates the boundary under pressure. Enterprise architecture keeps the pattern coherent across business units and platforms. Without explicit co-ownership, sovereignty becomes another security initiative rather than an institutional operating model.
8. **Use enterprise architecture for coherence, not gating.**
   Enterprise architecture should standardize reusable constraint patterns that platforms can enforce automatically. It should not become another approval body that teams learn to route around. EA succeeds when

production deployment, break-glass access, data export, identity delegation, vendor trust, and agent authority have reusable boundary patterns across the enterprise.

9. **Evolve SOC into D-SOC capability.**
   The SOC still detects and responds to attacks, but the D-SOC extends that mission by operating the decision boundary. It tracks boundary pressure, collapse integrity, trust inflation, reachability expansion, off-road execution, and constraint patch throughput. Restoration remains necessary, but structural closure becomes mandatory.
10. **Measure security leverage.**
   Track how many governed paved-road executions occur for every manual security intervention. The goal is to show that security workload does not grow linearly with business activity. As the paved road gains adoption, more work should flow through governed pathways without creating proportional growth in tickets, reviews, escalations, and analyst burden.
11. **Create an autonomy defensibility record.**
   For the first agentic or autonomous workflow, document the DAR category, authority boundary, curated actions, workflow states, collapse behavior, and unsafe trajectories made unreachable. The point is not to prove that every future agent behavior was predicted. The point is to prove that consequential authority was bounded.

These moves are modest enough to begin, but consequential enough to change what the security function is for. A dashboard doesn't become safer because it is more complete. An exception queue doesn't become governance because it is reviewed by more senior people. An incident process doesn't become mature because it restores service faster. Those activities are needed, but they are not the sovereign outcome. The sovereign outcome is a system in which consequential authority is visible, bounded, usable, measured, and designed to end.

That is the leadership shift. The CISO is no longer accountable only for protecting systems after they exist, or for approving work after intent has formed. CISOs become accountable for the conditions under which authority originates, expands, crosses boundaries, acts through humans or machines, and collapses when the mission ends. The work is not to own every decision. The work is to govern the paths through which consequential decisions become possible.

When those paths are governed well, security changes its relationship with the business. It stops asking the organization to slow down so oversight can keep pace. It builds the operating paths that enable organizations to move quickly without converting speed into trust inflation. The right path becomes pre-authorized. The dangerous path becomes unreachable. The temporary path ends by design.

That is the mandate of the sovereign CISO: help the organization move faster by making unsafe trajectories unreachable.

CHAPTER 11

Making the Transition: From Posture to Trajectory Governance

Trajectory governance is now on the table, and understanding the model isn't the hard part. Making it propagate through budgets, organizational charts, delivery pressure, executive incentives, and legacy systems is where the real work begins.

I have spent much of my career leading transformations across products, platforms, and services. I have seen change in legacy industrial environments, where constraints are physical, cultural, and deeply embedded. I have also seen change in high-tech environments, where constraints are less physical but just as real: incentives, funding models, delivery pressure, platform gaps, ownership ambiguity, and organizational fragmentation. Both types of environments resist change. But they resist for different reasons.

In legacy environments, leaders often have to move trust-first. People have learned how to keep it alive, where it breaks, which workarounds are safe enough, and which changes create operational danger. Before a leader can change the path, they must prove they understand what can't fail: uptime, continuity, customer commitments, and the credibility of the teams that have kept the system running. Trust creates the permission to change.

In fast-moving technology environments, leaders often have to move influence-first. Formal authority is rarely enough because teams move quickly, local incentives dominate, and the old path survives because it is faster, cheaper, and better aligned with how work gets done today. The leader doesn't unlock change by asking teams to believe in a new model. The leader unlocks change by making the new path more useful than the old one. In these environments, change spreads when the paved road becomes the easiest way to ship before enforcement arrives.

Trajectory governance isn't implemented by an announcement. A leader can't simply persuade the organization into a new security model. The more durable move is to change the organization's *choice architecture*: the defaults, workflows, platforms, incentives, approval paths, and operating rhythms that make one behavior easier than another. Culture follows those choices.

Many leaders spend years trying to change culture directly. That work often becomes abstract. Culture is a lagging indicator. It reflects the behaviors the environment has rewarded for a long time. If the off-road path is faster than the paved road, the culture will prefer the off-road path. If broad access is easier than bounded authority, the culture will normalize trust inflation. If exceptions are faster than product work, the culture will become exception-driven. If a security review arrives after intent has formed, the culture will experience security as rejection. The practical work of transformation is to change those conditions.

This chapter is the transition playbook that I've written for the leaders who have accepted the argument of the book and now have to make it real inside an organization that still has deadlines, legacy roles, old exception paths, audits, incidents, and business units that can't simply pause while security redesigns the operating model.

The goal is not to transform everything at once. The goal is to create the first undeniable proof. Make one unsafe trajectory unreachable. Make the governed path the path teams actually choose. Use the evidence to fund the next deletion. Then repeat until the model stops being a security initiative and becomes the way the organization moves.

Successful transformations follow a few recognizable patterns.

What Successful Transformations Do Differently

The successful transformations I have seen rarely begin as security mandates. They begin as *business problems* that security is able to help solve. A merger creates integration risk. A regulator creates urgency. A public incident changes executive appetite. A board-level scare makes the current operating model impossible to defend. The security program uses that moment to redefine its mandate. Without a forcing function, transformation often stalls because the organization can always choose to continue operating the way it already knows.

The second pattern is *sponsorship*. The CISO may lead the transformation, but the CISO rarely transforms the organization alone. There is usually a CEO, COO, CFO, product leader, or business executive who has decided that the security problem is also an operating problem. That sponsor gives the work legitimacy outside the security function. The CISO becomes the executor of a shared agenda, not the lone advocate trying to persuade the enterprise from the outside.

The third pattern is *metrics*. Successful leaders change the definition of progress before they change the program. If the organization still rewards ticket closure, tool deployment, audit completion, and exception throughput, the transformation will collapse back into the old model. The leader has to negotiate a different scoreboard early: adoption of the paved road, Mean Time to Authority, reduction in cheap trust, collapse success, and the number of unsafe trajectories actually removed. Metrics change what the organization protects, funds, and celebrates.

The fourth pattern is *embedding*. Successful security transformations do not win by building a stronger gate at the end of the workflow. They win by putting security capability inside the workflow: inside engineering platforms, product roadmaps, procurement motions, identity systems, deployment paths, data workflows, and operating cadences. The function dissolves into the work. Security remains accountable for the boundary, but the boundary is enforced through the systems people already use to get work done.

These patterns create the conditions for transformation, but they do not tell the leader where to act first. In every transformation I have led, one of the first practical challenges is visibility. This applies to security programs as well. The organization usually has dashboards, escalation stories, audit findings, and strong opinions, but it does not yet have a clear view of how work actually happens under pressure. Before a leader can choose the first unsafe trajectory to delete, form the right coalition, or negotiate the right scorecard, they need to see the operating reality: which paths teams use, where the old route still wins, where authority expands, and where the transformation is already being routed around.

That is why the transition starts with data fluency.

Build Data Fluency: See How Work Actually Happens

In my experience, most security programs are data-rich and insight-poor. They are good at counting vulnerabilities, issues, tickets, exceptions, policy violations, and remediation aging. They can tell you how many critical findings exist, how many teams missed an SLA, how many alerts were fired, and how many access reviews were completed. This data isn't enough for a transformation. Security transformation requires a different kind of data fluency.

Leaders need to understand how work actually happens inside an organization. Which paths do teams use under pressure? Which shortcuts are faster than the governed path? Where authority is being minted. Where it expands. Where it fails to collapse. Where the paved road is being adopted. Where it is being bypassed. Where trust inflation is still growing despite the transformation. his is where many security programs struggle.

Counting isn't enough. The data has to explain behavior. A transformation leader needs insights and explanations, not only on security issues but also on how the organization works and how people behave. That is why a data engineer or applied security data scientist becomes so valuable in this transition. They make organizational workflows observable enough that the leader can understand what is happening and decide what to change next.

The mistake is trying to build a complete security data platform before making the first move. That becomes another transformation inside the transformation. Start smaller. Pick the workflow selected in the first deletion: production deployment, break-glass access, privileged data export, vendor access, identity delegation, or agent tool use. Build just enough data fluency around that workflow to understand whether the paved road is winning or losing.

For the first workflow, the data spine should capture a few practical facts. It should show the actor, the action, the authority source, the path used, the time it took, the exception reason, the collapse result, and whether any new broad or persistent authority was created. This isn't a perfect model of the enterprise. It is a working picture of how one high-impact workflow behaves under real pressure.

A useful starting view can look like this:

Table 11.1: Data signals that describe a workflow

Data Signal	What It Reveals
Actor	Whether the workflow is initiated by a human, service account, pipeline, vendor, or agent.
Authority source	Whether authority came from the paved road, a standing role, a legacy token, inherited identity, or an exception.
Path used	Whether the team used the governed path, a manual ticket, a legacy workflow, or an off-road shortcut.
Latency	Whether the paved road is actually faster than the old path, especially at P90.
Exception reason	Whether the governed path is missing a real use case, too slow, too narrow, or not trusted during incidents.
Collapse result	Whether authority ended when the mission ended, or lingered as standing access.
Net-new cheap trust	Whether the organization is still creating broad roles, long-lived tokens, unmanaged trust edges, or inherited authority while claiming to transform.

This view gives the leader something more useful than just a security findings/ risk dashboard. It gives them a basis for action. If the paved road isn't being used, the data should show whether the issue is speed, coverage, usability, reliability, or habit. If exceptions are increasing, the data should show whether they are clustered around one missing capability. If collapse is failing, the data should show whether the workflow state is wrong, the cleanup is manual, or teams don't trust the access to remain available during real work.

The interpretation of data is as important as the collection. A high bypass rate usually means the paved road is missing a real use case or is too slow under pressure. A high paved-road failure rate means teams tried to use the governed path but couldn't complete work reliably. Low adoption despite availability usually means the old path is still easier or more familiar. Rising trust inflation during the transition means the old model is still manufacturing cheap trust faster than the new model can retire it. Adoption rising while trust inflation also rises means the paved road may be working, but the legacy path hasn't been deprecated.

Data fluency changes leadership behavior. The leader stops treating resistance as an attitude-only concern. The weekly review should stay brutally simple. For the selected workflow, review five things: paved-road usage versus off-road usage, Mean Time to Authority for both paths, exception reasons, collapse success, and net-new cheap trust created in the category. The purpose of the data is to decide what to ship, fix, deprecate, or enforce next.

A good weekly output is a short decision brief:

Table 11.2: Six fields of a weekly governance report

Field	Content
What changed this week	Adoption, bypasses, latency, collapse, and cheap-trust growth for the selected workflow.
Where the paved road lost	The team, use case, condition, or pressure point where the old path still won.
Why it lost	Missing action, slow authority, brittle collapse, unclear scope, reliability issue, or legacy convenience.
What must be shipped next	The product, platform, or control-plane change needed to make the governed path win.

Field	Content
What must be deprecated	The old convenience that keeps recreating trust inflation.
What must be frozen	Any net-new cheap trust that shouldn't be allowed while the category is under transition.

Good transformation data isn't the most complete data. It is the data that changes the next decision. For the first workflow, the leader should be able to say, with evidence, which path is winning, why it is winning, where authority is expanding, whether collapse is working, and what the coalition must ship next.

This is why data engineering and applied security data science should be brought into the transformation early. They make the transition visible before it becomes political. They show where the paved road is weak, where resistance is shaping, where trust inflation is hiding, and where the next deletion should happen. Without that capability, the transformation will be steered by anecdotes, escalation pressure, and the loudest story in the room.

The Hybrid Reality: Running Two Models Without Increasing Trust Inflation

You will be hybrid. It is the valid state of every real transformation. You will still scan, patch, respond, audit, and track exceptions. Meanwhile, you will also build and expand the boundary → curated action → collapse loop, which is delivered as a paved road. Posture activities remain necessary, but their role changes: they become hygiene, telemetry, and assurance rather than the primary mechanism for governing trajectories.

The objective of this phase is:

Operate both models; however, stop the old model from manufacturing new trust inflation while the new model ramps up.

Hybridity isn't dangerous merely because two models coexist. It can become dangerous, however, when the old route remains the easiest path under pressure. If legacy roles, broad tokens, ad hoc trust edges, and exception-heavy workflows continue to be faster than the paved road, your organization will keep widening trust while telling itself it's transforming. The result is the worst possible combination: more investment, more transformation effort, and a larger blast radius.

A practical way to manage hybrid is a sequence:

Subsidize → Instrument → Deprecate → Enforce

- **Subsidize:** Make the paved road faster than tickets for one high-impact workflow. If your "secure" path is slower, you will lose under deadline. Minimize your Mean Time to Authority.
- **Instrument:** Treat bypasses and overrides as telemetry. Don't moralize them. If the business goes off-road, it is signaling it has a product (i.e., paved road) gap.
- **Deprecate:** Use incremental friction to remove convenience from the legacy path deliberately. If running off-road remains one click away, it will remain the default under stress.
- **Enforce:** Enforcement comes last, not first. Only after the paved road is load-bearing do you harden the boundary so that unsafe trajectories become unreachable by construction.

The following two non-negotiables prevent hybridity from quietly increasing risk:

1. **Freeze cheap trust growth for the first Decision Authority Register (DAR) categories.** This is where the DAR becomes a transition tool, not just a governance artifact. For the first DAR categories, the rule is simple: Stop expanding the old trust supply, while you are trying to build the new path. It means you can't allow net-new standing administration roles, long-lived tokens without collapse semantics, or unmanaged cross-boundary trust edges. If you don't freeze the trust supply, you're paying to bail water while drilling new holes.
2. **Make collapse mandatory.** Hybridity is the period where deferred risk most easily becomes permanent infrastructure. Temporary access that doesn't collapse automatically isn't temporary; it's deferred risk.

A leadership artifact that can make this phase governable in practice is a *Hybrid Safety Contract,* which is one page in length and refreshed monthly. It states the following:

- What is structurally governed now (boundary + curated action + collapse)
- What remains posture-only temporarily (with an explicit sunset)
- What is frozen immediately (cheap trust rules)
- How exceptions work (exception = missing curated action to ship, or explicit time-boxed acceptance)

Hybridity succeeds when the overlap is controlled: The old model still runs, but it stops minting new trust debt faster than the new model can retire it.

Where to Start: Pick One Unsafe Trajectory and Delete It

This transition becomes real when one unsafe future becomes impossible. Not discouraged, flagged, or reviewed. It becomes *unreachable*. This is where many transformation programs fail. They start with breadth. They create a roadmap, a committee, a taxonomy, and a dashboard. They name workstreams, assign owners, and schedule steering meetings. Yet no one can point to a single class of failure that has actually disappeared from the organization's normal operating path.

You don't need breadth to start. *You need a first deletion.*

I have found the **Playing to Win** frame useful at this stage because it forces a practical choice. The leader has to decide where to play, how to win, what capabilities must be built, and what management system will sustain the change. In this context, "where to play" means the first consequential workflow. "How to win" means the unsafe trajectory that will be deleted. The required capabilities are the boundary, curated action, paved road, and collapse condition. The management system is the cadence that keeps adoption, bypasses, and trust inflation visible.

The right place to start is to pick workflow that is …

- High authority: Compromise or misuse creates meaningful blast radius.
- Frequent: The workflow happens often enough for adoption and bypass patterns to become visible.
- Under pressure: Teams already route around the ideal process when deadlines, incidents, or customer needs intensify.

- Producing trust inflation: The workflow already relies on broad reuse, lingering access, cross-boundary bridging, or inherited authority..

In most organizations, the first candidate usually appears in one of the following places:

- production deployment,
- break-glass access,
- privileged data export,
- high-authority service identity,
- vendor access,
- agent tool use,
- or an external action path that still depends on broad inherited authority.

The right first target is the one where deleting a single unsafe trajectory immediately changes daily behavior, reduces the blast radius, and proves that the new model is faster than the old one under real conditions.

For a production workflow, that first deletion might be simple and concrete:

Table 11.3: Starter workflows and the first trajectory to delete

Workflow	First Unsafe Trajectory to Delete	What Changes
Production deployment	Deploy to production without a validated artifact.	Deployment authority becomes tied to artifact provenance, service scope, environment, and workflow state.
Agent tool use	Let an agent inherit a broad human or service role for a narrow task.	Tool authority becomes task-scoped, brokered, and collapsed when the task ends.

The point isn't the specific category. The point is to choose one unsafe future and make it structurally unreachable. Then you should define "done" ruthlessly. A category isn't "governed" because it is simply documented. It is governed only when the following occurs:

- Boundary is enforceable in the control plane. The unsafe move is unreachable, not merely disallowed by policy.

- Curated action exists and is the normal interface for work. Teams can complete the workflow without off-road hacks.
- Collapse is automatic and reliable. Authority terminates when the mission ends.
- Paved road handles the great majority of executions for that workflow. The path isn't merely designed; it is adopted.

If you want to prevent self-deception, don't count the category until the paved road handles *at least 80% of executions* for that workflow.

This first deletion becomes the proof point for the transformation. It gives the organization a concrete before-and-after story: the unsafe trajectory used to be reachable, the old path used to win under pressure, and now the governed path carries most of the work. That proof changes the conversation. Leaders stop debating whether the model is interesting and start asking where the next deletion should happen.

Your first deletion should usually become your first fully governed DAR category. (Appendix F provides the operating format.) It's how you create momentum: One deletion proves the model, changes daily behavior, and creates a measurable speed dividend that makes the next deletion easier to fund.

Build the Coalition That Can Ship

Once you can see how work actually happens, the next challenge is building a coalition that can change it. Data fluency shows where the paved road is weak, where resistance is forming, where trust inflation is hiding, and where the next deletion should happen. But data by itself doesn't transform anything. It has to move into product decisions, platform work, operating cadence, and leadership tradeoffs.

I've seen many security transformations stall due to a lack of a coalition. The security team has the insight, the dashboard, and the executive support, but it doesn't have the machinery to change how work happens. The first deletion proves the model can work. Data fluency shows where the model is bending. The coalition determines whether the organization can repeat the pattern. A powerful sponsor outside security should be named explicitly. The sponsor does not own the security model, but they own the business pressure that makes the transformation non-optional. Without that sponsor, the paved road can become a security preference. With that sponsor, it becomes the operating path the business has agreed to fund and adopt.

For this transition to work, security has to become part of product and platform motions. The people who define the boundary, build the mechanism, analyze the data, drive adoption, and operate the pressure signals must all co-own the same outcome: make one unsafe trajectory unreachable while making the governed path easier to use than the legacy path.

The coalition could look something like this

Table 11.4: Owners and outputs during the transition

Owner	What They Own	Practical Output
Security leadership	The transformation target and governance boundary.	The DAR category, the unsafe trajectory to delete, the cheap-trust freeze, and the definition of done.
Applied security data science / data engineering	The evidence base.	Decision-space map, bypass analysis, paved-road adoption data, latency, collapse failures, and trust inflation signals.
Platform engineering	The machinery.	Authority foundry, curated action, identity issuance, workflow state, collapse mechanism, and control-plane enforcement.
Product management	Adoption.	Roadmap, usability, migration plan, Mean Time to Authority, user feedback, and deprecation of off-road paths.
Decision operations / D-SOC	The feedback loop.	Boundary pressure review, drift signals, exception patterns, incident-to-constraint-patch conversion, and operational follow-through.

Owner	What They Own	Practical Output
Enterprise architecture	Cross-domain coherence.	Reusable patterns that prevent local paved roads from creating enterprise-wide trust inflation.

This coalition works only when each owner has a real deliverable. Security leadership can't simply say what should be true. Platform engineering can't simply build a module and declare the work done. Product management can't just publish documentation. Data science can't just produce a dashboard. Decision operations can't just create another alert queue. Enterprise architecture can't simply review at the end. Each group has to own part of the path becoming real.

The most important practical rule is to **fund** the work as a platform capability, and not do it as a favor to security teams. If the authority foundry, curated action, collapse rule, or paved-road workflow is treated as something platform engineering does on the side, it will disappear when delivery pressure rises. Transformation becomes durable only when the secure path is part of the platform roadmap, resourced like other platform capabilities, and measured by adoption.

This is also where product management becomes non-negotiable. A paved road isn't adopted because it exists. It is adopted because it solves the user's problem better than the old path under real pressure. Product management owns the problem space: discovery, workflow research, UX, problem validation, feedback loops, adoption, and deprecation planning. The PM should understand why the old path is still winning, which users are bypassing, what jobs the paved road doesn't yet support, where the experience breaks during incidents or deadlines, and what must be deprecated once the governed path can carry the load. Without product ownership, the paved road becomes a technically correct artifact that no one uses. With product ownership, it becomes the default way work gets done.

Platform engineering and product management should be treated as partners, not downstream implementers. The security team brings the boundary and the risk logic. Platform engineering brings the mechanism. Product management makes the mechanism usable enough to win. Applied security data science shows whether it is actually winning. Decision operations keeps pressure visible after launch.

Enterprise architecture has a useful role, but it must not become a new gate. EA should help standardize patterns before every business unit invents its own paved road. It should make sure production deployment, break-glass access, data export, identity delegation, vendor access, and agent tool use don't get solved differently in every corner of the enterprise. EA adds value when it creates reusable patterns the platform can enforce automatically. It destroys value when it becomes another approval stop teams route around.

The most common failure is shared responsibility without shared accountability. The pattern I've seen work is: first, *name the shared outcome*. The cadence should review one shared scorecard, not five separate functional updates: adoption, Mean Time to Authority, bypasses, collapse success, and cheap-trust growth. In practice, shared accountability works when leaders stop asking whether each team completed its task and start asking whether the path changed. If the paved road exists but no one uses it, the coalition isn't done. If the boundary is defined but not enforceable, the coalition isn't done. If the path is adopted but cheap trust keeps growing, the coalition isn't done. Shared accountability becomes real when every owner can see the same outcome, the same evidence, and the same next paved road to ship

Cadence and Scoreboard That Make It Stick

The coalition works when there is a shared scorecard. In the absence of it, each function drifts back into its own definition of progress. And cadence is what turns that scorecard into behavior. It is the operating rhythm that converts evidence into product, platform, and paved roads. The cadence should stay simple: weekly, monthly, and quarterly.

In successful transformations, the scorecard changes early. If the old metrics remain dominant, the old behavior remains rational. The shared scorecard should therefore be negotiated before the program scales, not after the work is already underway.

In my experience, durable transformation requires rigor across both the *hard* and *soft* sides of change. The hard side is the constraint system itself of boundaries, foundries, paved roads, and collapse. The soft side is whether people trust it, use it, and stop routing around it under pressure. Both must move together if the transition is going to hold.

Weekly: Boundary Pressure Review

A weekly review is the shortest loop and tactical. Its purpose is to understand whether the paved road is winning under pressure. The coalition should review five signals every week:

Table 11.5: Signals that show the transition is working

Signal	What It Shows
Paved-road adoption	Whether teams are actually using the paved roads or still relying on the legacy route.
Mean Time to Authority	Whether bounded authority is faster than tickets, exceptions, or standing access.
Bypasses and overrides	Where teams went off-road and which pressure point caused it.
Collapse success	Whether authority ended when the mission ended or lingered as trust debt.
Net-new cheap trust	Whether new broad roles, long-lived tokens, unmanaged trust edges, or inherited agent authority were created.

This output isn't a compliance report. The output is a backlog. This weekly review keeps the coalition grounded in reality. Each review should end with the next product, platform, or control plane change needed to make the governed path win.

Monthly: Drift and Constraint Patch Review

The monthly review looks beyond individual bypasses and asks whether the system is drifting back toward old behavior. This is where the coalition identifies patterns that are too large for the weekly review.

Table 11.6: Drift patterns during the transition

Drift Pattern	Meaning
Temporary access becoming routine	Collapse isn't yet trusted, automatic, or tied to the workflow state.
Exception clusters	The paved road is missing a real use case or the boundary is modeled incorrectly.
Legacy roles surviving	The old path still carries operational dependency and hasn't been deprecated.
Trust inflation growth	The old model is still minting broad authority while the new model ramps up.
Agent or automation workarounds	New machine-speed workflows are being attached to inherited authority instead of governed pathways.

The monthly review turns repeated signals into structural correction. These reviews are where the organization decides which repeated signals are turning into patterns of drift and now require structural correction.

Quarterly: Expand DAR Coverage and Enforcement Readiness

The quarterly review is the institutionalization loop. It decides whether the selected category is ready to move from transition to enforcement, and it selects the next category to bring into the model.

For the current category, the quarterly review should assess four conditions:

Table 11.7: Enforcement readiness checklist

Condition	Enforcement Readiness Test
Boundary	The unsafe move is unreachable through the normal path, not merely discouraged by policy.
Curated action	Teams can complete real work through the paved road without off-road hacks.
Collapse	Authority reliably terminates when the mission ends.

Condition	Enforcement Readiness Test
Adoption	The paved road handles the great majority of executions, with 80% as a practical starting threshold.

The quarterly review should also add the next DAR category. The quarterly cadence prevents transformation from becoming endless strategy. Every quarter should produce either measurable stronger enforcement in an existing category or a new category entering the transition sequence.

Leading Indicators That the Transformation Is Working

The strongest signs of success usually appear before the metrics look perfect. They show up in language, behavior, and budget.

Table 11.8: Leading indicators of paved-road adoption

Leading Indicator	What It Means
Teams ask for paved-road access.	Paved roads are becoming useful enough that teams want these before enforcement.
Postmortems change language.	Teams stop blaming human error and start naming missing curated actions, brittle collapse, or off-road paths.
Platform engineering supports the funding request.	The paved road is now seen as platform capability, not a security favor.
Product managers ask about adoption and UX.	The transformation has moved from policy to product.
Executives ask about blast radius and trust inflation.	Leadership is beginning to think in terms of reachable consequence, not only compliance status.
Exception reviews produce backlog items.	The organization is converting friction into product and platform work.

The organization starts speaking differently, funding differently, and interpreting problems differently. That is the point at which the change is moving from project to institution.

What Cadence Prevents

The above cadence prevents three common failure modes.

First, it prevents the coalition from declaring victory too early. A paved road that exists but isn't used isn't a transformation. It is an option. Adoption keeps the coalition honest.

Second, it prevents the old model from silently expanding underneath the new one. Trust inflation growth indicates whether legacy roles, broad tokens, inherited identities, and unmanaged trust edges are still increasing, even as the organization claims to be moving toward bounded authority.

Third, it prevents resistance from becoming political. Bypasses, exceptions, and overrides are reviewed as operating signals. They show where the paved road is missing coverage, losing on speed, or failing under pressure.

The cadence isn't there to keep everyone busy. It is there to make structural progress habitual. Weekly reviews keep the coalition close to reality. Monthly reviews convert repeated pressure into constraint patches. Quarterly reviews expand DAR coverage and decide when enforcement is justified.

Transformation needs to become durable before they become complete. This coalition rhythm is how transformation stops depending on heroic leadership energy. It becomes part of how the organization runs.

The Leader's Pivot: From "Chief No Officer" to "Chief Architect"

This final shift is personal. Many leaders are implicitly rewarded for heroism (e.g., a firefight, an incident bridge, a decisive escalation). However, the trajectory-governed model rewards something else: *architectural closure.* It rewards the work of making a class of failure unreachable, so that the next incident never needs that same intervention in the first place. It's a different kind of leadership, and for many CISOs, it is the hardest shift.

This shift requires three pivots:

- **An ego shift:** Stop equating value with intervention. Your goal isn't to be needed at every moment of danger. Your goal is to help build a system where intervention becomes rarer because the boundary holds.

- **Language shift:** Security leaders have to stop narrating everything as generic risk. Risk is often too abstract to drive action. Paths are concrete. A leader who can say, "This unsafe sequence is reachable; we will collapse it" creates clarity and momentum.
- **Promise shift:** Stop promising that vigilance and response will keep pace with your system's speed. That promise is no longer credible. A better promise is that the business will be able to move faster, because the most dangerous trajectories are bounded by construction and not left to constant vigilance.

The shift from "chief no officer" to "chief architect" is ultimately a move from pessimism to agency. In the old model, the CISO is a professional pessimist, constantly reminding the board that the world is dangerous and a breach is inevitable. It is a wearying way to lead. In the sovereign model, you become a builder. You're no longer just managing a list of problems; you're designing a system that's fundamentally more capable than the one you inherited. The organization will stop auditing memories and start governing trajectories. It will stop hoping people will remember and start building environments where the safe path is the normal path, and unsafe trajectories simply don't exist.

How the Leader's Week Changes

This pivot becomes real when the leader's attention changes, as shown in the following table:

Table 11.9: Posture-era vs. transition leader focus

Posture-Era Leader Focus	Transition Leader Focus
Reviews exception tickets and decides whether this one is acceptable.	Reviews whether the exception reveals a missing paved-road capability.
Escalates findings because the dashboard says they are overdue.	Prioritizes findings that reveal reachable unsafe paths or authority expansion.
Measures whether teams completed remediation work.	Measures whether the system changed and the unsafe path became harder or impossible to reach.

Posture-Era Leader Focus	Transition Leader Focus
Uses incidents to justify more monitoring or staffing.	Uses incidents to drive constraint patches and structural closure.
Treats bypasses as compliance failures.	Treats bypasses as evidence that the old path is still easier, faster, or more reliable.
Reports control counts, audit status, and backlog.	Reports DAR coverage, speed dividend, trust inflation trend, and structural closure.
Pushes teams to follow the process.	Funds product and platform work that makes the right process easier to follow.
Is rewarded for catching unsafe behavior.	Is rewarded for making classes of unsafe behavior unreachable.

The leader's focus isn't on becoming less operational. The leader's focus is on becoming more operational at a higher layer. Instead of supervising every dangerous move, you're shaping the environment in which those moves can or cannot exist.

A Monday Morning Ritual

Most of us have spent our careers in the posture-based world, so the shift to trajectory governance will feel uncomfortable at first. You will feel the urge to return to the familiar comfort of the critical alert, the exception queue, and the compliance checklist. *Resist it.*

The complexity of modern systems has outpaced our ability to supervise them. We can no longer "watch" our way to safety. We must "build" our way there.

On Monday morning, don't look at your vulnerability dashboard first. Instead, look at your organization's reachability ratio. Ask yourself: How many of our people are currently "off-road"? How much "cheap trust" did we manufacture over the weekend? Then pick one category of authority, or one high-stakes lever, or one unsafe trajectory, and *delete it.* Then, build the paved road, collapse the standing access, weld the boundary shut, and repeat it.

Looking back, the journey from posture to trajectory governance can be understood as a progression of questions. Security traditionally begins with the state. State asks: **Where are we?** A posture report, an asset inventory, a vulnerability scan, or an access review attempts to answer that question. These views remain useful. They describe the environment as it exists today.

Behavior asks a different question: **How did we get here?** It reveals movement. It shows how permissions were used, how workflows operated, how systems interacted, and how events unfolded. It provides more context than state, but it still describes outcomes after movement has already begun.

A deeper question follows: **What decisions brought us here?** Every meaningful trajectory is ultimately composed of decisions. Access was granted. Trust was extended. Authority was delegated. A workflow was approved. An exception was accepted. A deployment path was created. A vendor was integrated. These decisions shape the path the system ultimately takes.

But even decisions do not fully explain risk. The most important question is: **What futures are possible from here?** This question sits at the heart of trajectory governance. A system's future is shaped by its authority, workflows, interfaces, trust relationships, defaults, incentives, and constraints. Together, they determine which paths are available, which paths are difficult, and which paths are impossible.

This progression reflects the broader argument of the book:

State describes what exists.

Behavior explains how the system moved.

Decisions explain which path was chosen.

Trajectory governance determines which paths can be chosen at all.

For decades, security has focused primarily on the first question and increasingly on the second. Modern systems, automation, and autonomy force us to confront the third and fourth.

The goal is not to observe unsafe futures more quickly. The goal is to make unsafe futures harder, narrower, or impossible to reach in the first place. The goal was never "perfect security." The goal is a sovereign enterprise: A system that knows its own boundaries, honors its own constraints, and empowers its people to move with the speed that the modern world demands.

The tools are in your hands, and the trajectory is yours to shape.

Go build.

APPENDIX A

First Principles for Securing Decision Systems

Traditional security was built for systems whose behavior could be largely understood as instruction execution. Agentic AI systems add a different layer: They interpret context, evaluate options, and select actions under uncertainty.

When systems decide under uncertainty, operate with expanding authority, and pursue objectives across time, the foundations of security must change.

This appendix is my first-principles approach that is required to secure such systems.

0. Agency changes the security problem.

Automation follows a script. Agency follows an objective.

A script can be reviewed line by line. Its path is fixed in advance. An agent is different, however. It interprets, adapts, selects, and moves. It doesn't merely execute a predefined sequence. It navigates toward an outcome. That distinction changes the security problem.

You can audit a script. You must constrain an agent.

Implication: The rise of agency shifts security from verifying execution to governing motion.

1. AI systems are decision systems.

AI systems don't simply produce outputs. They select actions from a space of possibilities based on context, training, and objectives.

Risk isn't a property of any single output. It emerges from how decisions accumulate, interact, and unfold over time.

Implication: Security must govern the trajectory of decisions, not just the state of the system.

2. The attack surface is behavioral.

Attackers no longer exploit only code paths or vulnerabilities. They exploit how systems interpret instructions, weigh context, and prioritize actions.

A system can be technically secure and still behave unsafely, because the compromise is no longer only in the code. It is in the path the system is induced to take.

Implication: Securing inputs and outputs is insufficient. The reasoning process itself becomes part of the attack surface.

3. Authority expands faster than responsibility.

AI systems are rapidly granted access to data, tools, and workflows. They read broadly, act programmatically, and operate across systems.

Governance rarely keeps pace with this expansion. Capability arrives first. Constraint arrives later, if at all.

Implication: Security must explicitly define and constrain decision authority before autonomy is introduced.

4. Context is both power and risk.

AI systems derive their usefulness from context. The same context defines their exposure. Context can be incomplete, misleading, stale, or adversarially manipulated. It doesn't simply inform decisions. It reshapes the range of decisions the system believes are reasonable.

Implication: Context must be scoped, validated, and governed as a first-class security surface.

5. The snapshot fails in non-deterministic systems.

Traditional security relies on the snapshot—a point-in-time inspection of a static state. It assumes that if the state is acceptable at the moment of inspection, risk is acceptably managed. That assumption collapses in AI systems.

In a probabilistic system, a system's state is fluid, reasoning is dynamic, and risk lives in motion. A snapshot of an AI system is a map of a ghost; by the time you have analyzed it, the reasoning path has already shifted.

Implication: Security controls must operate on trajectories, not snapshots, and must tolerate uncertainty rather than pretend it can be eliminated.

6. Models are not trusted components.

Models can hallucinate, misinterpret, overgeneralize, and fail under adversarial pressure. They don't possess judgment. They simulate it, treating model behavior as trustworthy, because it is useful is a category error.

Implication: Models must be treated as untrusted components. Their outputs may inform decisions, but they must never be the final authority over consequential action.

7. Alignment is not security.

Alignment attempts to shape what a system should do. Security defines what a system *can* do. These aren't the same. A system can be aligned in intent and still unsafe in operation. It can follow the spirit of a goal while also crossing an unacceptable boundary on the way there.

Implication: Security requires enforceable constraints. Alignment alone isn't a control plane.

8. Risk aggregates at trust boundaries.

The most significant failures occur at the boundaries: between user and system, between system and tools, and between systems themselves. These boundaries are the moments where context is translated into action and where authority is transferred, expanded, or amplified.

Implication: Trust boundaries, not individual components, are the primary surface that must be secured.

9. Observability must explain trajectories, not just outcomes.

Traditional logs capture what happened. They don't explain why it happened, what context shaped it, or how the system arrived there.

In decision systems, outcomes without reasoning are insufficient for control. An executed action is the end of a path, not the path itself.

Implication: Security requires visibility into context, reasoning, and decision trajectories, not just events and results.

10. Governance must precede motion.

Autonomy amplifies both capability and failure. Once a system is granted the agency to move, weak design choices compound at machine speed.

Control cannot be retrofitted after deployment. By then, the universe of possible action has already been defined.

Implication: Decision authority must be architected in the foundry before a system is granted the agency to move.

What These Principles Reject

These principles represent a formal evolution of security for an autonomous world. They explicitly reject:

The snapshot fallacy: The belief that risk can be managed by inspecting static states.

The compliance mirage: The idea that alignment with a policy manual is a substitute for hardcoded constraints.

The oversight trap: The assumption that "human-in-the-loop" can scale to the speed of agentic decision-making.

The detection tax: The reliance on fixing things after the decision has already been executed.

In an autonomous world, we don't need faster scanners. We need better-designed universes.

AI security isn't the protection of models, data, or infrastructure in isolation. It is the governance of how systems decide, what authority they hold, and which trajectories they are allowed to take over time.

Security doesn't need to happen earlier. It needs to happen differently.

APPENDIX B

Trajectory Governance vs. Whitelisting

Many security teams will instinctively translate trajectory governance into familiar language: "So, it's just allowlists but more." That translation is understandable and wrong. Trajectory governance isn't a new allowlist strategy. It is a different unit of control: It's not a permission list, but a rule about valid evolution of state over time.

B.1 The Core Distinction

Whitelisting governs actions. It answers: "Is this action/tool/resource allowed?"

Trajectory governance governs decision space. It answers: "What next states and actions are reachable from here, by whom, under what authority, through which interfaces, and with what follow-on consequences?"

B.2 Why "More Layers of Whitelisting" Isn't the Same Thing

A common response is to add layered allowlisting: identity, tool, parameters, resource targets. Layered allowlisting is still action-scoped validation. It catches more, but it catches the same kind of thing.

Layered whitelisting answers the following:

- Who can call this tool?
- With which parameters?
- Against which resources?

Trajectory governance answers the following:

- Is this step valid *from this state*?
- Does this sequence preserve the intended workflow graph?

A single tool call may be individually "allowed" and still be dangerous when combined with the previous step. The difference becomes obvious in any environment that has seen multistep attacks over a time period.

B.3 Example: RunProcess Is Whitelisted vs. Graph-Governed

Consider an agentic tool environment (Model Context Protocol [MCP] as an example, but the pattern generalizes to cloud control planes). The system exposes tools like the following:

- ReadFile
- RunProcess
- Possibly network tools, deployment tools, etc.

Whitelisting Model

You define allowedTools = ["readFile", "runProcess"]. When the model calls a tool, you check whether the tool is on that allowlist. If yes, you run it (maybe after schema validation).

This is static capability control. It answers "Is this tool allowed in this environment?"

It does *not* answer "Is this tool allowed in this moment, in this trajectory?"

Trajectory Governance Model

You define a state machine/transition graph. For example:

- State: Code Review
  - Allowed transitions: ReadFile, SuggestEdit
- State: Build Validation
  - Allowed transitions: RunProcess limited to build/test
- State: Deployment
  - Allowed transitions: RunProcess limited to deploy routines

Now, RunProcess isn't "globally allowed." It's *contextual:* It's allowed only in certain states and only after specific preceding transitions.

So, the check is no longer "Is RunProcess allowed?" It becomes "Is runProcess a valid transition from the current state?"

If the agent is in *Code Review*, RunProcess is unreachable, not filtered, or not "blocked after evaluation," it's simply not a valid move in the system. That is the mental model shift.

B.4 Sequences Matter More Than Steps

In agentic environments, models chain tools creatively. If the system exposes a set of powerful tools and only allowlists them, an agent can legally assemble them into an unintended outcome.

Consider a customer-support agent with three permitted tools: LookupCustomer, SendEmail, and CreateExport. A user message arrives: "*Please send me a copy of my account history for tax purposes.*" The agent does exactly what it's allowed to do:

- Step 1: LookupCustomer: retrieves the customer's full account record, including payment methods and historical transactions.
- Step 2: CreateExport: generates a CSV of the retrieved data.
- Step 3: SendEmail: delivers the CSV to the address provided in the user message.

Each step is allowed. Each tool is on the allowlist. Schema validation passes. The agent has followed instructions. The address it sent to, however, is attacker-controlled (supplied through a prompt-injection), and the data the customer asked for includes records the customer was never entitled to see at this depth. The harm is not in any individual call. The harm is in the composition: the export step was valid in isolation but should not have been reachable from this conversational state, addressed to this destination, with this scope.

Trajectory governance constrains composition by making the workflow graph explicitly define:

- Which tools may follow which tools
- Which states permit which transitions
- Which transitions are terminal
- (Optionally) whether cycles are even possible

In the same scenario under trajectory governance, the agent's state after LookupCustomer would not permit CreateExport as a valid transition unless the customer's verified contact channel matches the destination, the scope is bounded to the requesting user's own data, and the export action has a defined terminal state. The harmful sequence becomes structurally unreachable — not because each step is blocked, but because the *transition* between them is not a defined edge.

Security failures in tool environments are often trajectory failures, not single-step violations.

B.5 Graph Consistency as the Enforcement Mechanism

A system is considered to be *graph consistent* if every step it takes corresponds to a valid edge in the approved transition graph:

- No illegal jumps
- No skipped states
- No undefined paths
- No actions after the workflow has ended

The system is not merely checking whether an action is allowed. It is checking whether the action belongs at this point in the work. A valid action in the wrong place is still wrong.

Consider temporary production access. The approved path might be: request submitted → owner approves → JIT access minted → session used → ticket closed → access collapses. Adding a user to a production-access group may be a valid action in general. It is not valid before approval, outside the approved service, with broader scope than requested, or after the ticket has closed. Graph consistency makes those distinctions enforceable.

This differs from allowlisting. An allowlist says, "this action is permitted." Graph consistency says, "this action is permitted only as the next valid move in this workflow." That keeps enforcement focused on sequence rather than a growing list of allowed tools, APIs, roles, and exceptions.

It also makes investigation easier. The log reads like a path: requested → approved → minted → used → collapsed. When the path is broken, the missing or unexpected step is visible.

B.6 What Threat Classes Become Structurally Eliminable (and What Don't)

Trajectory governance is powerful, but it isn't magical. It shifts some categories from "mitigate" to "eliminate," while others remain traditional security concerns.

Often structurally eliminable (when encoded correctly):

- **Unauthorized tool invocation/capability pivot** (tool not in the reachable graph for the current state)
- **Cross-tool/cross-server chaining** (no defined transition across servers/tools)
- **Arbitrary target injection** (parameters/targets derived from prior state outputs rather than free-form strings)

Partially reducible (not fully eliminable) items:

- **Prompt injection:** You can eliminate escalation into non-reachable tools, but manipulation within the allowed path can remain.

Items not addressed by trajectory governance (still required):

- **Transport-level attacks (man-in-the-middle [MITM], replay, and spoofing)**: Still needs strong transport identity and cryptography
- **Supply chain/compromised tool server:** If the component itself is malicious, graph rules won't save you.

B.7 A Practical Implementation Pattern You Can Reuse Across Domains

Even outside MCP, the implementation pattern is consistent:

1. **Define states** that reflect intent (i.e., review, build, deploy; debug, migrate, remediate; investigate, contain, recover).
2. **Define allowed transitions** per state (tools/actions), with bounded parameter domains.
3. **Enforce transition validity** at the control plane layer (where intent becomes authority).
4. **Record state transitions** as the primary audit log (not just actions).
5. **Measure drift** as deviation pressure on the graph through the following:
    - Repeated override attempts
    - Attempts to "skip" stages
    - Pressure to add new transitions
    - Growth in reachable authority

This is how trajectory governance becomes operational and measurable, and why it's not reducible to "a better allowlist."

B.8 Trajectory Risk Reduction Matrix

When you apply trajectory governance to an agent/tool architecture, some threat classes move from **mitigate → eliminate** (structurally), while others remain traditional controls.

Table B.1: Agent threat classes: traditional vs. trajectory

Threat Class (Agent/Tool Systems)	Traditional Control	With Trajectory Governance
Prompt injection → tool escalation	Filters/post-processing	**Mostly eliminated** (escalation becomes unreachable if tools aren't valid from the current state)
Unauthorized tool invocation	Allowlists	**Fully eliminated** (tool not part of allowed transitions)
Cross-server/cross-tool pivot	Policy/manual review	**Fully eliminated** (no valid transition across servers/tools unless defined)
Infinite tool loops	Rate limits	**Fully eliminated** (terminal states, monotonic progression, cycle-free graphs)
Parameter/target injection	Schema validation	**Mostly eliminated** (parameters derived from prior state outputs, no arbitrary new targets)
MITM/replay/downgrade	TLS/mTLS	**Not addressed** (below decision space)
Supply chain compromise	SBOM/signing	**Not addressed** (if the tool server is malicious, the graph can't save you)
Overprivileged downstream creds	IAM	**Partially addressed** authority foundry's job; see the *Authority Foundry* section in Chapter 8.)

Two Rules of Thumb to Avoid "Whitelisting Relapse"

Rule 1: If your control grows as a list, you're enumerating, ask what transition you're trying to make unreachable.

When you catch yourself adding entries ("allow one more tool," "one more domain," "one more exception"), pause and restate the real goal as a transition constraint:

- "This tool may only run after Build Validation has completed."
- "External network transitions aren't part of this workflow."

Restating a real goal is how you move from action permissioning to workflow integrity.

Rule 2: If you rely on filters and parameter validation, you're still governing steps—govern the sequences by enforcing state + progression.

Schema validation is useful, but it can't see "legal steps, illegal outcome." Trajectory governance adds the missing structure like the following:

- Explicit states
- Monotonic progression (no skipping gates)
- Terminal states (no wandering)
- Bounded parameter domains tied to prior outputs (no arbitrary new targets)

This is true beyond agents. Any system that operates by composing valid actions under pressure, such as CI/CD pipelines, cloud control planes, and automated remediation, will see the same failure pattern. Governance that can't constrain trajectories will always degrade into an exception economy.

APPENDIX C

Constraint Patching Toolkit

Let's start with an example scenario: A team needs production data for debugging, analytics, or model evaluation. The supported export path is slow or unclear, so they move the data through an ad hoc route: direct query, copied token, one-off script, or legacy integration.

Constraint Root Cause Analysis (RCA)

1. **Outcome:** Sensitive data left the environment through an unstructured export path.
2. **Trajectory:** Need for data identified → paved road perceived as slow or incomplete → operator used legacy or bespoke pathway → data moved without purpose-bound authority or terminal state.
3. **Unreachable point:** The system shouldn't have made an ad hoc export the fastest operational path for legitimate work.
4. **Patch location:** Interface + Transition
5. **Patch:**
    - Ship a brokered export action that requires explicit dataset, purpose, destination, and duration.
    - Make the export authority pathway purpose-bound and time-bounded.

- Add sink controls and terminal completion state (i.e., collapse).
   - Increase friction on bespoke export routes, so the paved road becomes the dominant path.

6. **Regression proof:**
   - Negative transition test: Export is rejected without a declared purpose and sink.
   - Path-dominance test: Supported export path works end to end for the common use case, and the bespoke route is measurably harder.
   - Terminal-state test: Export authority collapses when the transfer window closes.
7. **Adoption metric:**
   - Percentage of exports through brokered pathway
   - Drop in ad hoc export events
   - Drop in exceptions for "temporary" export access
   - Improved Mean Time to Authority for legitimate export requests

The goal isn't to eliminate data movement. It is to eliminate unstructured data movement as the easiest default behavior.

Now, I'll generalize this example and create a template for your toolkit.

C.1 Constraint RCA Template

Use this template when closing any meaningful incident or repeated near miss.

1. **Outcome:** What happened that we never want again?
2. **Trajectory:** What sequence of transitions made it reachable?
3. **Unreachable point:** Which step should have been unreachable by construction?
4. **Patch location:** Authority/Interface/Transition
5. **Patch:** What change deletes or structurally shrinks this trajectory?
6. **Regression proof:** How do we prove this stays unreachable?
7. **Adoption metric:** What signal confirms that the system moved onto the paved road?

C.2 Patch Location Cheat Sheet

- **Authority patch (issuer/compiler):** Tighten minting rules, scope, duration, and prerequisites.
- **Interface patch (paved road update):** Add/upgrade curated actions so that teams stop improvising.

- **Transition patch (progression rules):** Involve gates, prerequisites, terminal collapse, "no self-modifying governance"

C.3 Minimum Viable Regression Proofs

Pick one that matches the patch type:

- **Issuer test:** "Authority cannot be minted without a prerequisite state."
- **Negative transition test:** "This move is rejected outside the paved roads."
- **Terminal-state test:** "Elevated state collapses when the mission ends."
- **Path dominance test:** "Paved roads works end-to-end; an off-road path is measurably harder."

C.4 Post-Patch Boundary Metrics

- Override rate (Did bypass drop?)
- Authority duration and scope (Did trust inflation stop?)
- Terminal-state integrity (Does collapse happen reliably?)
- Denied-transition clustering (The spike during rollout is normal; sustained clustering needs analysis.)

C.5 Weak Patch vs. Strong Patch

A useful way to keep postmortems honest is to test whether the fix changes the system or only the conversation around it.

Table C.1: Weak-patch vs. strong-patch pairs

Weak Patch	Strong Patch
Add a reminder.	Remove the unsafe move from the normal workflow.
Tighten review wording.	Change minting rules, interface, or progression so that the path is structurally narrower.
Add another alert.	Bind the collapse or prerequisite state into the control plane.
Require more approvals.	Ship a paved-road action that covers the real-use case safely.
Track exceptions better.	Make the off-road route slower, more visible, and less ambient.

APPENDIX D

Authorizing Agents: Why the Existing Models Fail

Every authorization model in production today was designed for humans.

That sentence should unsettle you more than it does. When a human analyst opens a dashboard, there is an implicit context surrounding that action:

- A reason for access
- Professional norms constraining behavior
- A manager who will eventually review the work
- A career that depends on not doing something catastrophically unsafe

Authorization systems have never had to encode any of that context explicitly, because it was already present in the human sitting in front of the keyboard. AI agents have none of it. They have no career. They have no professional norms. They have no felt sense of what counts as catastrophically unsafe. What they have is a prompt, a set of retrieved context, and whatever the model weights produce at inference time.

Additionally, AI agents act at machine speed. This means that when we hand an agent our existing authorization credentials: OAuth tokens, role memberships, or relationship tuples, we aren't extending our security model to a new kind of user. We are exposing the gap that was always there: The gap between what our authorization systems govern and what actually determines whether an action is safe.

The Default: Delegated Identity

The most common pattern today for authorizing AI agents is the simplest one: Give the agent the user's credentials. The agent inherits your OAuth tokens, session, and permissions. It can do whatever you can do. In most enterprise deployments, this isn't even a deliberate design choice, it's just what happens when you plug an LLM into an existing API layer and don't think too hard about the authorization question.

The appeal is obvious. There is no new infrastructure to build or no new permission model to design. The agent acts as you, and whatever controls already exist for you apply to it.

The problem is equally obvious once you state it clearly: The agent's purpose is narrow, but your authority is broad. A scheduling agent doesn't need access to your financial systems. A code-review agent doesn't need access to your human resource (HR) records. Delegated identity conflates who you are with what the agent should be allowed to decide.

This is the posture problem restated at the authorization layer. Delegated identity describes a static fact like "This agent holds these credentials" and treats it as sufficient for security. It says nothing about what the agent is trying to accomplish, what decisions it should be making, or what trajectory it is on. It is configuration masquerading as governance.

Worse, it makes the blast radius uncontrollable. When an agent operating under a delegated identity gets prompt-injected, hallucinates a dangerous action, or simply misinterprets its instructions, the blast radius is everything you can do. It's not everything the agent was supposed to do, rather it's everything the human principal is authorized to do. The failure is bound by human identity, not by the agent's task.

Practitioners will recognize the pattern. We have seen it before in service accounts with domain-admin privileges, in API keys with full-access scopes, and in shared credentials that persist long after the original purpose has expired. Allowing delegated identity for agents is the same mistake at a new layer of abstraction, but it's moving faster and with less visibility.

The Instinct: Role-Based Access Control

The natural response, for anyone trained in enterprise security, is to reach for role-based access control (RBAC). Give the agent its own identity. Assign it a role. Scope the role narrowly. Taking these actions is better than delegating them. At least the agent is no longer operating with your full authority, however, RBAC breaks down in ways that matter.

RBAC answers a specific question: Given who this entity is, what resources can it access? It maps identities to roles and roles to permissions on resources. For human users operating within organizational structures, this works tolerably well. Humans act slowly, within social context and with professional accountability. The coarse-grained nature of the roles is compensated by the fine-grained judgment of the humans who hold them.

Agents lack every one of those compensating factors. Additionally, RBAC's limitations, which are manageable for humans, become structural failures for agents.

RBAC governs resources, not decisions.

RBAC grants access to things—read this table, call this API, write to this bucket. But agent risk doesn't live at the resource layer. An agent with read access to your customer relationship management (CRM) and write access to your email isn't dangerous because of either permission individually. The agent is dangerous because of the decision sequence it can compose: Read customer data, draft a message, and send it externally.

RBAC cannot express the constraint: "You may read the CRM, and you may draft emails, but you may not chain those two actions into an outbound communication without human approval." It governs nouns. Agent risk lives within actions, actions that are composed over time.

RBAC is static; agent behavior is trajectory-dependent.

A role is the same at 9 A.M. and 3 A.M. in the morning. It is the same whether the agent has made 1 API call or 500. It doesn't change based on what the agent has already done in the current session.

But agent risk is deeply context dependent. An agent that has already read sensitive data should have fewer options for its next action, not the same options. The permission set should contract or expand, based on the decision path the agent has taken so far. RBAC has no concept of this. It is a snapshot model applied to a problem that requires trajectory awareness.

RBAC assumes stable mapping has occurred between identity and intent.

When you assign a human the analyst role, there is a reasonable assumption that their intent aligns with the role's purpose. The analyst wants to analyze. They have professional context, social accountability, and career incentives that keep their behavior roughly within the boundaries the role implies.

An agent's intent is whatever the prompt, retrieved context, and model weights produce at inference time. That intent can shift mid-session due to a prompt injection, hallucination, or unexpected context. A static role assignment cannot track or respond to intent drift. It was never designed to do so.

RBAC doesn't handle agent-to-agent composition.

Agents will increasingly invoke other agents. In an RBAC model, you check the downstream agent's role against the target resource. However, the actual governance question is different: "Should this agent be allowed to act on behalf of that agent's current task, given the decisions already made and the trajectory so far?" That is a delegation-chain problem, not a role-membership problem. RBAC has no vocabulary for it.

None of this means that RBAC is useless in an agentic context. You still need resource-level controls as a backstop. But treating RBAC as the authorization model for agents is like treating firewall rules as your security strategy; it governs the wrong layer of abstraction.

The Sophisticated Answer: Relationship-Based Authorization

Practitioners who have moved beyond RBAC will point to a more expressive model: relationship-based access control, most prominently implemented in Google's Zanzibar system and its open-source descendants. This is worth taking seriously, because it represents the most mature version of structural authorization available today, and it still falls short.

Zanzibar's core innovation is elegant. Instead of asking "What role does this entity have," it asks, "What is this entity's relationship to this object?" Permissions are computed by traversing a graph of relationships like the following: A user belongs to a team, the team owns folder, the folder contains document, therefore, a user can access document. This gives you things RBAC cannot, like transitive permission chains, composable delegation, and fine-grained structural reasoning about who relates to what.

For agents, this is a genuine improvement. You can model delegation as a relationship edge: User A granted Agent X access to Resource Y. You can model an agent-to-agent invocation as another edge. The graph can, in principle, represent multihop delegation chains that RBAC can't.

Zanzibar's model answers a fundamentally structural question: Does a path exist in the relationship graph that grants this permission? However this question isn't sufficient for governing-agent behavior.

The graph authorizes access, not action sequences.

Zanzibar evaluates point-in-time checks: Does Entity E have Relationship R to Object O such that Permission P is granted? Each check is independent.

However, the dangerous thing an agent does is rarely a single access. It is a chain of access composed into a consequential action. Zanzibar will authorize each step in the chain independently and has no mechanism to evaluate the sequence as a trajectory. It will say yes to reading customer records. It will say yes to drafting an email. It will say yes to sending that email externally. It has no way to see that these three individually authorized operations together constitute a data exfiltration.

The relationship graph is structural, not temporal.

Zanzibar's graph represents what organizational structure, ownership, membership, delegation is. It doesn't represent what has happened. It cannot express: "This agent already accessed sensitive data in this session, so its permission to perform external actions should now be restricted."

The graph is a map of the world, not a log of the journey through it. For humans, this limitation is usually acceptable, because human decision-making is slow and contextual. For agents acting at machine speed, though, the journey is the risk.

Delegation in Zanzibar is structural, not scoped.

You can represent "User A delegated to Agent X" in Zanzibar. But what you actually need to express is something closer to "User A delegated to Agent X for the purpose of scheduling meetings this week, using only the work calendar, with no access to email content."

That is a task-scoped, time-bounded, conditionally constrained delegation. Zanzibar's tuple model (entity, relation, subject) doesn't have a natural place to put those constraints. You either create an explosion of narrow synthetic relationships, which defeats the elegance of the model, or you bolt constraints on as a separate enforcement layer, which means Zanzibar is no longer your authorization system in any meaningful sense. It is the storage backend for a system that doesn't yet exist.

No native concept of attenuation.

Agents need the idea that authorization depletes. An agent should be able to make *n* API calls, spend up to a certain dollar amount, or take actions up to a cumulative risk budget before it must stop and reauthorize.

Zanzibar's relationships are binary: They either exist or they don't. There is no concept of a permission that weakens with use or a trust boundary that narrows as the session progresses. You could approximate this with external state that mutates the graph in real time, but at that point you're fighting the model rather than using it.

What the Field Actually Needs

If neither RBAC nor Zanzibar is sufficient, what would a correct authorization model for agents look like? The question matters less as a design specification and more as a way of clarifying what we are actually governing.

A structurally sound agent authorization system would have several properties that no existing model provides, including:

- The permission set is defined by the *task*, not by the entity's identity or organizational position.
- It narrows dynamically as decisions are made and context changes.
- It governs *sequences of actions,* not individual resource access.
- It gravitates toward restriction under uncertainty. If the system cannot determine whether the next action is within scope, the default is to stop and escalate, not to permit.

This is decision-space governance applied to authorization. The boundary isn't "What can this entity access?", but "What decisions should be allowed, given what has already happened and what we know about the intended trajectory?" The model must be trajectory-aware, task-scoped, temporally sensitive, and capable of attenuation. None of which is to say that RBAC and relationship-based models have become irrelevant. You still need resource-level controls, and Zanzibar-style systems are excellent infrastructure for that layer. But they become the floor, not the ceiling. They enforce structural invariants—this agent cannot touch resources outside its structural authorization—while a decision-governance layer on top enforces behavioral constraints: what action sequences are sanctioned, how the permission set evolves with each step, and when the system must pause and involve a human.

That second layer, the decision-governance layer, doesn't exist yet as a coherent system in the industry. Pieces of it appear in various places: approval workflows, spend limits, and anomaly detection. But no one has assembled them into a principled model of agent authorization grounded in the idea that risk is a property of decisions made over time.

This isn't an accident. Building that layer requires to model the decision space ahead of time—to specify, before the agent acts, what trajectories are sanctioned, and what boundaries should hold. That is exactly the capability most organizations lack, because they have spent decades optimizing for posture assessment rather than for decision governance. The tooling gap for agent authorization is a symptom of the broader conceptual gap.

Practitioners don't need to build the perfect authorization system before deploying agents. But they do need to see clearly what the existing models are and aren't governing. Delegated identity governs nothing. RBAC governs resources. Zanzibar governs relationships. None of them governs decisions, and decisions are where the risk lives.

APPENDIX E

From Insight to Institutionalization: How Security Ideas Spread in Organizations

Strategic ideas rarely move from insight to adoption instantly. Inside engineering-driven organizations, ideas typically mature through several stages as understanding, credibility, and operational evidence accumulate. Research on innovation adoption describes this engineering pattern in different ways. Communication theorist and sociologist Everett Rogers, in his book Diffusion of Innovations, outlines the progression from knowledge to persuasion, decision, implementation, and confirmation. Author of Crossing the Chasm, Geoffrey Moore's work on technology adoption reflects a similar logic: What feels compelling to early believers still requires proof before it becomes organizationally normal.

Within security organizations, these dynamics often manifest as a gap between the leader's conceptual clarity and the organization's need for operational proof.

Leaders often reach the insight stage months before the broader organization. What feels obvious to the leader may still feel theoretical to engineers, product partners, or executive peers. That gap isn't a failure of intelligence. It's a normal feature of organizational change.

The role of leadership during transformation is, therefore, not to prematurely finalize the solution, but to *advance the idea from one maturity stage to the next* by introducing the right artifacts like problem framing, mental models, workflows, and prototypes at the appropriate moment.

Table E.1: Stages of an idea spreading through an org

Stage	Leader's Mental State	Organization's Mental State	Typical Risk	Best Leader Actions to Move to Next Stage
1. Intuition	"Something is off; there is a pattern here."	Not yet aware of the issue	The idea is vague and difficult to communicate.	1. Surface the problem clearly using data, customer stories, or operational friction. 2. Frame the stakes. Describe why the current model is insufficient.
2. Insight	"I see the pattern and	Skeptical but curious	The idea sounds	1. Introduce a simple mental

Stage	Leader's Mental State	Organization's Mental State	Typical Risk	Best Leader Actions to Move to Next Stage
	the potential direction."		conceptual or abstract.	model or principle. 2. Connect the idea directly to observable operational problems.
3. Articulation	"I can clearly explain the new model."	Interesting, but theoretical	Engineers struggle to visualize implementation.	1. Translate the idea into a workflow or "day-in-the-life" scenario. 2. Show a before/after comparison that demonstrates practical value.
4. Demonstration	"Here is proof that it can work."	Interest and engagement increase	Limited trust may exist until the results are visible.	1. Build a small prototype or pilot. 2. Demonstrate how the idea solves a real operational problem.

Stage	Leader's Mental State	Organization's Mental State	Typical Risk	Best Leader Actions to Move to Next Stage
5. Institutionalization	"This is clearly the right model."	Broad alignment and adoption	Risk of stagnation or premature standardization exists.	1. Embed the idea into tools, workflows, and roadmaps. 2. Reinforce success through internal case studies and metrics.

The point isn't to move through these stages ceremonially. It is to recognize that different organizational moments require different forms of proof:

- A leader who keeps offering mental models when the organization needs workflow examples will sound abstract.
- A leader who keeps offering pilots when the organization needs institutional ownership will look tactical but incomplete.

Leaders advocating shifts in security models, such as moving from posture management to trajectory governance, often find themselves several stages ahead of the broader organization. The most effective path forward isn't persuasion through argument but progression through artifacts, including problem narratives, operational workflows, and working prototypes.

Table E.2: Artifacts that move an idea between stages

Stage Transition	Useful Artifact
Intuition → Insight	Problem narrative
Insight → Articulation	Mental model or conceptual framework
Articulation → Demonstration	Workflow examples

Stage Transition	Useful Artifact
Demonstration → Institutionalization	Prototype or operational tool

A useful way to operationalize this appendix is to think in terms of an artifact ladder:

- **Problem narrative:** What friction, incident pattern, or recurring exposure shows that the old model is failing?
- **Mental model:** What new explanation makes the problem legible? (In this book: snapshot fallacy, decision space, trajectory, and trust inflation.)
- **Workflow example:** What does the new model look like in one real path? (Deployment, break-glass, data export, agent external action)
- **Operational proof:** What unsafe future has now become unreachable? What is the first deletion?
- **Institutional scaffolding:** What decision authority register (DAR) category, cadence, budget line, ownership model, and scoreboard make the pattern repeatable?

The Institutionalization Test

An idea is not institutionalized because people can repeat its language. It is institutionalized when the organization …

- funds it,
- measures it,
- builds around it, and
- begins to experience it as the normal path rather than the exceptional one.

For the trajectory governance model, that means …

- a paved road is faster than the old route,
- the first governed DAR categories are real,
- resistance shows up as product feedback instead of political drama, and
- teams stop asking whether security will approve the move and start asking which supported path they should use.

It is the point where a security idea becomes an organizational reality.

APPENDIX F

The Decision Authority Register

A Decision Authority Register (DAR) is a control-plane governance artifact. It's a trajectory register for the handful of consequential decisions, where intent becomes authority expressed as what's unreachable (the curated path) and how authority collapses.

It isn't a policy checklist nor is it a list of "approved roles." Its job is to make unsafe trajectories structurally unreachable and to slow (or reverse) trust inflation, by preventing authority from expanding by convenience and persisting by default.

A consequential decision tells you what must be governed. The boundary tells you what must be unreachable. Curated actions are the supported moves that make that road usable.

F.1 The DAR Format (Decision → Boundary → Path → Collapse)

Table F.1: Example DAR entries

Consequential Decision	Bounded Transition (the Boundary)	Curated Action (the Path)	Collapse Trigger (the End State)
Production deployment	No deployment is reachable without a signed provenance attestation minted by CI for this artifact.	promote_to_production (artifact_digest)	**One-time deployment authority collapses immediately after completion** (hard TTL ≤ 15 minutes as a backstop)
Database (DB) debugging	Direct production DB access is unreachable except within an active P1 incident workflow.	request_diagnostic_view (service, incident_id)	**Authority collapses on incident close** (session termination + token revocation), with a max Time-To-Live [TTL] cap
Data replication	Replication is reachable only for approved classification pairs and approved destinations (e.g., Class-A → Class-B only).	replicate_cross_account (dataset_id, destination_id)	**Cross-boundary trust collapses on completion** (trust edge removed automatically)
Identity delegation	A subtask token cannot exceed the parent token's	mint_subtask_token (parent_token, task_scope)	**Automatic revocation on parent**

Consequential Decision	Bounded Transition (the Boundary)	Curated Action (the Path)	Collapse Trigger (the End State)
	scope (no privilege escalation by delegation).		**termination** (and on task completion)
Security control modification	Disabling Web Application Firewall (WAF)/firewall controls is unreachable without a dual-signature break-glass token tied to an incident/change window.	emergency_bypass_open (control_id, window)	**Automatic revert** at end of window (≤ 60 minutes) + post-window verification

F.2 How to Use the DAR (Minimum Viable Guidance)

- **Keep it small.** Start with 5–12 decision categories (i.e., deploy, privileged access, data movement, identity minting/delegation, external side effects, and security control changes).
- **Boundaries must be "unreachable unless …."** If the boundary reads like a guideline, it will degrade into posture management.
- **Curated actions are the interface.** If the "path" is "edit IAM/run raw API/use a static admin role," the system is still governing by memory and review.
- **Collapse is non-negotiable.** A system that can mint authority but cannot reliably collapse it will accumulate trust edges over time (trust inflation by design).

- **Treat bypass as a signal.** Repeated overrides aren't considered "noncompliance." They are evidence that the curated action is missing, misfit, or too slow for reality.

F.3 Anti-Patterns: The "Whitelisting Relapse"

When drafting your DAR, avoid these failure modes that drag the program back into posture management:

1. **The "Checklist" Entry:** If an entry says, "Ensure all users have MFA," it belongs in a policy document, not the DAR. A DAR entry must define *transition logic* (e.g., "step-up MFA is the mint-time trigger required for the foundry to issue this class of token").
2. **The "Static Role" Entry:** If the curated action is just a link to a permanent identity and access management (IAM) role, you haven't governed the trajectory. You've labeled a legacy problem. Curated actions should be workflows/capabilities that mint bounded, time-scoped authority with explicit collapse.
3. **The "Ghost Boundary"**
   Defining a boundary your infrastructure cannot enforce creates false confidence. If the boundary isn't construction-ready, it belongs on the risk register as an unmitigated risk until the foundry/interlock exists.

F.4 When a DAR Category Counts as Structurally Governed

A category counts as structurally governed only when …

- the boundary is enforceable in the control plane,
- a curated action exists and is the normal interface for doing the work,
- collapse is automatic and reliable, and
- as the program matures, the paved road handles the great majority of executions for that category.

APPENDIX G

Identifying the High-Entropy Surfaces: The Big 5

If the scope of "governing decisions" feels infinite, it is usually because the organization hasn't identified its *high-entropy pathways.* These are the specific decision points that, if left unconstrained, create the most reachable future risk.

In many modern environments, a small number of decision categories shape a disproportionate share of structural risk. These five categories are a practical starting point for the *Decision Authority Register (DAR).* The goal isn't to claim that every organization has exactly the same five categories.

The goal is to start where authority, speed, and blast radius usually converge.

1. State Change Authority

- **Scope:** Deployments, rollbacks, configuration flips, and feature flags
- **Sovereign Shift:** We stop trying to "approve" code. We govern the *Paved Roads.*
- **Constraint:** No change can reach production, unless it is the result of an attested monotonic progression (e.g., Build Scan Test Sign).

2. Privileged Diagnostic Access

- **Scope:** Break-glass, incident debugging, and direct administrative actions
- **Sovereign Shift:** We eliminate "standing admin." We treat debugging as a temporary transition.
- **Constraint:** High-privilege authority can only be minted in response to a verified incident and must physically collapse (i.e., session termination) the moment the mission ends.

3. Data Movement and Replication

- **Scope:** Exports, cross-account replication, training-set generation, and third-party sharing
- **Sovereign Shift:** We stop relying on data loss prevention (DLP) filters. We govern the destination control plane.
- **Constraint:** Data can only be moved via a *curated action* that enforces destination-bucket encryption, classification matching, and mandatory TTLs on the copy.

4. Identity Minting and Delegation

- **Scope:** Issuing tokens, assuming roles, and establishing cross-account trust
- **Sovereign Shift:** We move from "Static IAM" to an authority foundry.
- **Constraint:** No identity can "self-mint" broader authority. All expansion must happen through a *constraint compiler* that scopes the new token to a specific, time-bounded intent.

5. External Impact Surface

- **Scope:** Sending messages, issuing refunds, changing user entitlements, or disabling security controls
- **Sovereign Shift:** We govern the *Blast Radius of Side Effects.*
- **Constraint:** These actions are unreachable for any agent or user, unless they are triggered through an interface that enforces per-transaction "circuit breakers" and budget limits.

APPENDIX H

Reachability Ratio Toolkit

Reachability Ratio measures how much of a defined environment an identity, workload, integration, workflow, or agent can affect if its authority is misused. It is a practical blast-radius metric. It does not ask only whether an entity has a risky permission. It asks how much future damage becomes reachable through that entity's current authority, trust relationships, network position, workflow access, and control-plane reach.

A reachability ratio isn't a single ratio for the entire organization. A single average would be a vanity metric that hides dangerous outliers. Instead, the reachability ratio is reported as a distribution or a heatmap across your different domains. Think of it like a weather map: You don't care about the "average temperature of the country." You care about where the heat, the blast radius, is concentrated.

Reachability Ratio = Reachable Impact Surface / Defined Impact Surface

For example, if a CI/CD runner can affect 100 production assets out of a defined production universe of 1,000 assets, its Reachability Ratio is **10%**. If a vendor integration can affect 500 assets out of that same universe, its

Reachability Ratio is **50%**. If a purpose-bound deployment token can affect only one service out of 800, its Reachability Ratio is **0.125%**.

Lower is better. A lower Reachability Ratio means the entity has a narrower blast radius and a more bounded future.

What Counts as Reachable

Reachability should not mean only network reachability. In trajectory governance, an asset is reachable if the entity can materially affect it through any meaningful path.

That path may include:

- Identity reach: assuming roles, minting tokens, using delegated permissions, or crossing trust boundaries.
- Network reach: connecting to systems, services, APIs, databases, or internal endpoints.
- Data reach: reading, exporting, modifying, replicating, or deleting data.
- Deployment reach: changing artifacts, deployment definitions, configuration, release state, or rollback behavior.
- Control-plane reach: altering identity, policy, secrets, workflow state, infrastructure, or administrative settings.
- External-effect reach: sending messages, issuing refunds, changing entitlements, triggering transactions, or taking actions outside the organization.

A read-only path to one dataset is not equivalent to administrative control over a deployment system. Mature versions of the metric should therefore weight reachability by sensitivity, authority type, environment, and consequence. But the first version does not need to be perfect. It only needs to make the concentrated blast radius visible.

What the Operator Scoreboard Looks Like

Here is how the operator scoreboard actually appears in CISO's internal report. It turns thousands of connections into a few actionable "pressure points."

Table H.1: Reachability ratios across domains

Domain	Reachability Ratio	Status	Operational Signal
Public Web Tier	1:800	Green	Constraints are holding; the blast radius is localized
CI/CD Pipelines	1:10	Yellow	**Boundary Pressure:** Pipelines have broad access; need to move to CI-minted tokens
Third-Party SaaS	1:2	Red	**Trust Inflation:** These integrations can "see" almost everything; an immediate decision authority register (DAR) priority
AI "Research" Lab	1:5	Red	**Off-Road Execution:** Teams are bypassing the Foundry to give agents "full admin" for speed.

The Ideal: The 1:*n* Ratio

In a perfect sovereign system, the numerator is 1:

- **1:** The specific purpose-bound asset the entity is authorized to touch right now.
- ***n*:** The total number of assets in the environment.
- **Result:** The result is a "needle-thin" trajectory. If the entity is compromised, the blast radius is limited to that one single "1."

The Reality: The X:*n* Ratio

When trust inflation or boundary pressure kicks in, that numerator (X) starts to climb.

- **X:** The number of assets an entity *could* reach if it went rogue, because of "standing" permissions or broad network access.
- **N:** The total assets
- **The Signal:** If your CI/CD pipeline has a ratio of 100:1,000, your numerator is 100. You are telling the operators: "This one entity has the 'vision' to destroy 100 things."

Why This Is Better Than a List of Findings

Reachability Ratio should feed directly into the Decision Authority Register and the constraint-patching process.

A high Reachability Ratio asks:

- Which unsafe trajectory is still reachable?
- Which authority path made it reachable?
- Is the issue a missing curated action, a broad role, a standing token, a trust edge, a raw control-plane interface, or a failed collapse condition?
- What constraint patch would make the same reachability narrower next time?

For example:

The third-party SaaS integration can affect 50% of sensitive repositories and metadata stores. This is not merely a vendor-risk finding. It is a trust-inflation problem. The constraint patch is to replace the standing integration token with a scoped pathway that grants repository-level access only for the approved purpose and expires when the integration task ends.

That is a more useful leadership conversation than "go fix 400 findings."

APPENDIX I

The Security Product Manager's Manifesto

Product Managers (PM) come in various flavors. One day, I might decide to share much more detail about what I've learned as a product manager. Although I didn't want to leave the security product managers' manifesto untouched. In the sovereign model, a security product manager doesn't manage "compliance." They manage adoption. If the business bypasses a constraint, it isn't a "disciplinary issue"; it is a product failure.

The following five tenets define the mindset and mechanics required to ship structural governance.

1. The Speed Dividend Is the Primary Metric

Security traditionally adds friction. A security PM's mission is to subtract it.

- **The Goal:** The paved road must be measurably faster than the off-road path.
- **The Action:** If a ticket takes three days and a paved road takes three minutes, you have won. If the paved road takes four days, your product is "broken" and trust inflation will inevitably rise.

2. Solve for Jobs-to-be-Done, Not Checklists

Developers don't wake up wanting to "be secure"; they wake up wanting to ship code.

- **The Shift:** Don't ask: "How do we secure S3?" Ask: "What is the developer trying to do when they touch storage?"
- **The Action:** Build curated actions that fulfill the developer's intent (e.g., "request encrypted bucket") as a single atomic unit of work.

3. Treat Bypass as Telemetry, Not Defiance

When a team circumvented a boundary, they provided you with the most valuable product data available:

- **The Mindset:** A bypass is a feature request.
- **The Action:** Map every exception or shortcut back to a gap in the paved roads. Is the road too narrow? Is it too slow? Is the collapse mechanism breaking their workflow? Use this telemetry to prioritize the next engineering sprint.

4. Optimize for the Default State

Human behavior follows the path of least resistance:

- **The Strategy:** Use choice architecture to make the safest path the easy path.
- **The Action:** Preconfigure the foundry with sane defaults. If a developer must change five settings to be secure, they may miss one. If they must change five settings to be insecure, they must do so intentionally.

5. Manage Deprecation Debt

The hardest part of product management isn't shipping new features; it is killing old, unsafe ones:

- **The Strategy:** Follow the Subsidize → Instrument → Deprecate → Enforce sequence.
- **The Action:** You cannot enforce a boundary until the paved road can handle 80% of the load. Your job is to drive that 80% adoption through superior user experience (UX), not through "chief no officer" mandates.

The Security PM's Monday Morning Scorecard

- **Mean Time to Authority (MTTA):** How long does it take a user to become authorized on the paved road versus the legacy ticket path?

- **Adoption Rate:** What percentage of new projects start on the paved road?
- **Friction Coefficient:** How many manual steps (i.e., approvals/slack pings) remain in the curated action?
- **Target:** Zero.

Glossary

This book uses some familiar security terms in a specific way and introduces several new terms to describe how risk forms in dynamic, automated, and authority-driven systems. The purpose of this glossary isn't to define every cybersecurity acronym or tool category. It is to give the reader a working vocabulary for the book's main argument: security must move beyond describing current posture and toward governing the trajectories systems are allowed to take.

Some terms in this glossary are common but have specific meanings here, such as **posture**, **control plane**, **guardrail**, **authority**, and **reachability**. Other terms are specific to the model developed in this book, such as **trajectory governance**, **decision space**, **authority foundry**, **trust inflation**, **paved road**, **automatic collapse**, and **Decision Authority Register**.

The glossary is organized by conceptual categories rather than as one long alphabetical list. The terms in the book aren't isolated definitions; they form a system. The reader should be able to see how the mental model terms connect to authority terms, how authority terms connect to design primitives, how design primitives connect to metrics, and how those metrics connect to leadership and transformation.

Category 1: Core Mental Model Terms

Conditional Risk

Risk whose severity depends on context, sequence, reachability, authority, and surrounding conditions rather than on the finding alone.

A vulnerability, permission, or configuration may be dangerous only if it sits near privileged authority, connects to a sensitive workflow, or can combine with other conditions to create broader consequence. This is why fixed severity labels often underrepresent dynamic risk.

Critical Mass

The point at which individually reasonable grants, permissions, exceptions, integrations, or trust relationships accumulate into a materially different form of risk.

Critical mass is reached when a set of "acceptable" decisions creates enough reach to dissolve boundaries, expand blast radius, or make an unsafe trajectory reachable.

Decision Space

The set of moves available inside an environment. It includes what users, systems, workflows, identities, software, suppliers, and agents can do, reach, inherit, influence, or trigger next.

Decision space is broader than inventory and broader than reachability. It captures not only what can be accessed, but what future states and actions become possible through authority, workflow state, interfaces, and trust relationships.

Posture

Security's current-state map of an environment: assets, configurations, vulnerabilities, permissions, exposures, policies, and controls.

Posture is useful because it shows what exists now. It becomes dangerous when mistaken for a full model of what can happen next. The book doesn't argue against posture; it argues that posture alone is incomplete.

Reachability

What can be accessed, touched, influenced, or reached from a given starting point.

Reachability is an important part of decision-space reasoning, but it is narrower than decision space. Reachability shows that a path exists; decision space explains what that path enables, how it composes with other paths, and which futures it makes possible.

Reachable Future

A future state or outcome that can be reached through the current configuration of authority, interfaces, workflows, permissions, and trust relationships.

A reachable future may not be intended, expected, or frequently used. Attackers, automation, and agents can explore paths the organization didn't mean to operationalize.

Snapshot Fallacy

The belief that a sufficiently accurate description of the present state is enough to explain future risk.

The snapshot fallacy doesn't mean snapshots are wrong. It means they are being asked to explain too much. In dynamic systems, risk often forms through authority changes, workflow composition, trust expansion, and accumulated decisions that a single snapshot can't adequately explain.

Trajectory

The path an **identity**, workflow, system, actor, or agent takes through decision space over time.

A trajectory is formed when allowed actions compose into a sequence. Each step may be valid in isolation, while the sequence creates a future the organization didn't intend to permit.

Trajectory Governance

The practice of shaping which sequences of action are possible within a system.

Trajectory governance is concerned not only with whether an individual action is allowed, but whether a dangerous path should be reachable at all. It extends security from observing the current state to governing reachable futures.

Category 2: Authority and Control-Plane Terms

Ambient Authority

Authority that is available broadly to an actor or system without being specifically minted for a task, workflow, purpose, or terminal state.

Ambient authority often appears as inherited roles, standing permissions, broad service accounts, or user tokens reused by automation or agents. It is dangerous because the actor can use more authority than the immediate task requires.

Authority Accumulation

The gradual concentration of permissions, trust relationships, dependencies, or capabilities inside one identity, workflow, system, or integration.

Authority accumulation usually happens through reasonable local decisions. A role gets one more permission. A workflow gains one more environment. A service account is reused for one more use case. Over time, the accumulated authority creates a broader trajectory than the organization intended.

Authority Binding

The moment authority is attached to an identity, workflow, system, piece of code, pipeline, agent, or service.

Authority binding is where risk often begins. A token is minted, a role is assumed, a cross-account trust edge is created, or a pipeline is given deployment power. By the time posture tools observe the result, the more important decision may already have occurred: the system has been authorized to move along a path.

Authority Collapse

The structural ending of authority when the mission, workflow, task, or state that justified it has ended.

Collapse is stronger than cleanup. It doesn't rely on someone remembering to revoke access later. Authority disappears because the workflow reached its terminal state: the incident closed, the deployment completed, the export ended, or the agent task finished.

Authority Foundry

A control-plane mechanism that mints bounded authority only when the right workflow state, purpose, scope, and conditions are present.

An authority foundry doesn't simply hand out persistent permissions. It creates short-lived authority for a specific use: debugging a service during an active incident, deploying a verified artifact, exporting an approved dataset, or allowing an agent to call a tool for a bounded task.

The foundry is where persistent policy becomes temporary capability. It translates the organization's rules into usable authority that is narrow, contextual, and designed to collapse.

Authority Minting

The creation of usable authority for a specific workflow, task, purpose, state, or action.

Authority minting is different from assigning a persistent permission. A persistent permission says what an identity may be allowed to do in general. Minted authority creates a narrower capability for a particular moment: a specific incident, deployment, data export, agent task, or operational workflow.

In a trajectory-governed system, minted authority should be purpose-bound, scoped, short-lived, and tied to workflow state. It should collapse when the mission ends.

Authority Surface

The set of identities, workflows, systems, tools, integrations, and control-plane paths through which authority can be exercised.

The authority surface is different from the attack surface. The attack surface asks where an attacker might enter. The authority surface asks what an actor can do once inside and how far authority can propagate.

Bounded Authority

Authority that is limited by purpose, scope, workflow state, time, resource, environment, or terminal condition.

Bounded authority is the opposite of broad standing power. It lets teams move quickly without giving them more reach than the task requires. In a trajectory-governed system, authority should be bounded before it is used, not only reviewed after it causes concern.

Bridging Identity

An identity that connects systems, environments, accounts, datasets, or trust boundaries that were intended to remain meaningfully separate.

A bridging identity often forms through authority accumulation. It may start as a narrow role or service account and grow through legitimate additions until it becomes a path across boundaries. The identity didn't bypass the boundary; it became the bridge that weakened it.

Cheap Trust

Broad, persistent, reusable authority that is easy to grant and hard to unwind.

Cheap trust includes standing admin roles, long-lived tokens, unmanaged trust edges, broad service accounts, vendor credentials, and inherited agent authority. It is cheap because the system makes it easy to widen trust in the moment. It becomes expensive later because teams come to depend on it.

Control Plane

The layer of systems that author authority, define valid state, configure resources, orchestrate workflows, and determine which future actions are reachable.

Control planes include identity systems, CI/CD orchestrators, cloud management APIs, Kubernetes API servers, service-mesh policy layers, SaaS administration surfaces, policy engines, and agent tool brokers. Execution planes run workloads. Control planes decide what workloads exist, what authority they hold, how they connect, and how they can change.

Permission

A durable rule or configuration that defines what an identity, role, system, or actor is allowed to do. Permissions often live in roles, policies, access-control lists, service accounts, group memberships, or resource policies. They are usually more persistent than the immediate task that required them. A permission describes what could be exercised if the identity holds or inherits it.

In this book, permissions are not the same as minted authority. A permission is a standing rule. Minted authority is a bounded capability created for a specific purpose, state, scope, and time window.

Purpose-Bound Authority

Authority created for a specific purpose, task, workflow, dataset, service, incident, artifact, or action.

Purpose-bound authority limits what an actor can do to the reason the authority was granted. It prevents a narrow task from inheriting broad power. This is especially important for automation and agents, where inherited roles can turn a narrow workflow into a broad blast radius.

Purpose-bound authority is usually minted, scoped, and short-lived. It shouldn't persist after the purpose has ended.

Trust Boundary

A boundary that defines where trust is expected to change: between environments, accounts, organizations, vendors, networks, identity domains, data classifications, or operational contexts.

Trust boundaries weaken when identities, integrations, or workflows bridge them without corresponding structural constraints. In this book, trust boundaries matter because unsafe trajectories often form when authority crosses them too easily.

Trust Edge

A relationship that allows trust or authority to pass from one system, identity, account, organization, or workflow to another.

Examples include cross-account roles, federated identities, vendor API tokens, service-to-service trust, CI/CD deployment roles, and agent tool permissions. Trust edges are powerful because they create paths through decision space.

Trust Inflation

The expansion of trust through inherited roles, persistent access, reusable tokens, cross-boundary relationships, vendor integrations, and broad identities without a corresponding increase in structural constraint.

Trust inflation devalues the isolation boundaries an organization designed. It is rarely malicious. It is usually the result of rational local decisions that make trust easier to expand than to constrain. Over time, trust inflation increases the number of unsafe trajectories the system can reach.

Category 3: Trajectory-Governance Primitives

Automatic Collapse

The built-in termination of authority when the workflow, task, incident, deployment, export, or mission that justified it ends.

Automatic collapse is stronger than manual cleanup. It doesn't depend on a person remembering to revoke access or close a session. The system itself removes the authority when the terminal condition is reached.

Boundary

A defined limit around where authority may originate, what it may affect, under what conditions it may be used, and where it must stop.

A boundary isn't only a policy statement. In trajectory governance, a boundary should be enforceable in the control plane. It determines which actions, states, workflows, or transitions are valid.

Constraint Design

The practice of shaping systems so unsafe trajectories are harder, narrower, or impossible to form.

Constraint design doesn't mean adding more checkpoints around an unchanged system. It means changing the system so certain unsafe moves are no longer normal moves the system can express.

Constraint Patch

A structural change shipped after an incident, near miss, bypass, or repeated exception reveals that an unsafe trajectory is still reachable.

A constraint patch may tighten a boundary, add a curated action, change an authority foundry rule, add a workflow state, remove an off-road path, or bind collapse to a terminal state. It is the difference between learning about a failure and changing the system that made the failure possible.

Sequence Integrity

The property that a workflow can move only through valid states in a valid order.

Many unsafe outcomes aren't created by a single forbidden action. They are created when valid actions occur in the wrong order, skip prerequisites, or connect into a path the organization never intended.

Structural Closure

The condition reached when a known unsafe trajectory has been removed, narrowed, or bounded by a system change.

Structural closure is different from closing a ticket. A ticket may be closed while the path remains reachable. Structural closure means the system itself changed so the same class of unsafe trajectory is harder or impossible to repeat.

Structural Constraint

A control embedded into architecture, interface design, authority boundaries, workflow states, or valid transitions so an unsafe path isn't available as a normal option.

A structural constraint doesn't simply detect or reject a bad move late. It changes what the system can express. This is the primary mechanism by which trajectory governance becomes real.

Terminal State

The workflow state that marks the end of the mission, task, incident, deployment, export, or agent action.

Authority shouldn't persist after the reason for it has ended. In a well-designed system, terminal state triggers collapse.

See also: Automatic Collapse; Terminal-State Integrity; Stateful Progression; Authority Collapse.

Trajectory Regression Test

A test that verifies a previously deleted unsafe trajectory remains unreachable after a change to policy, workflow, paved-road configuration, or authority foundry rules.

Trajectory regression tests are especially important when securing the paved road itself. They test safety claims such as: production deployment without validated artifact remains unreachable, privileged access without an active incident remains unreachable, and authority can't persist after terminal state.

Category 4: Paved-Road and Platform Terms

Choice Architecture

The design of defaults, workflows, interfaces, incentives, and friction so that some choices become easier, more natural, or more likely than others.

In this book, choice architecture explains why security succeeds or fails inside real organizations. Teams don't choose paths only because policies say so. They choose the paths that are fastest, clearest, most reliable, and most aligned with how work actually gets done.

Curated Action

A supported, intent-level action that lets a user, workflow, platform, or agent perform useful work without exposing raw control-plane power.

A curated action turns a broad capability into a safer interface. Instead of giving a pipeline broad deployment authority, the platform may expose a curated action such as promote_to_production(artifact_digest). Instead of giving an engineer standing production access, the platform may expose request_diagnostic_view(service, incident_id).

Off-Road Path

A workaround, legacy route, manual process, broad role, exception path, or custom workflow that bypasses the supported path.

Off-road paths often survive because they are faster, more familiar, or more reliable under pressure than paved roads. They aren't always signs of bad behavior. They are often evidence that the paved road lacks real capability, has poor usability, or hasn't yet earned adoption.

Paved Road

A supported path for doing high-impact work safely, quickly, and consistently.

A paved road packages the boundary, curated action, authority foundry, workflow progression, and collapse condition into a usable path that teams can rely on. It isn't merely documentation or a recommendation. It is the way safe work becomes easy enough to become the default.

Speed Dividend

The business benefit is created when the governed path becomes faster, clearer, and more reliable than the legacy path.

The speed dividend appears when teams can obtain bounded authority, deploy safely, access data, debug incidents, or use agents through paved roads faster than they could through tickets, exceptions, or standing access. It is how security becomes an accelerant rather than a checkpoint.

Category 5: Governance Artifacts and Metrics

Consequential Decision

A decision whose execution materially changes what the system can do next.

Consequential decisions create or expand authority, cross trust boundaries, move sensitive data, cause irreversible external effects, or change the controls that govern the system itself. Examples include production deployment, break-glass access, privileged data export, identity delegation, vendor access, and agent tool use.

Consequential decisions matter because they are the decisions that belong in the Decision Authority Register. They shouldn't be governed only through informal judgment, late review, or recurring exceptions.

Control-Plane Constitution

The set of structural rules that determine where authority can originate, how it can expand, which paths it can take, and how it must collapse.

The control-plane constitution is a leadership concept. It describes what the sovereign CISO is accountable for: not approving every decision, but ensuring that the systems governing high-impact decisions enforce the right boundaries by construction.

Modern security failures often emerge where authority is authored, widened, reused, and preserved. A control-plane constitution makes those authority rules explicit.

Decision Authority Coverage

The portion of the organization's consequential decision surface that is governed structurally.

A Decision Authority Register category counts as covered only when it is complete in practice, not merely documented. The boundary must be enforceable, the curated action must exist, collapse must be reliable, and the paved road must handle the great majority of real executions.

A decision category isn't structurally governed because it appears in a register. It is governed when the system actually enforces the path and teams actually use it.

Decision Authority Register

A governance artifact that records the consequential decisions an organization has committed to governing structurally.

A Decision Authority Register, or DAR, isn't a risk register. A risk register tracks what the world might do to the organization. A DAR tracks what the organization allows its systems to do. It names the decisions where authority must be bounded: production changes, privileged access, data movement, identity delegation, external side effects, vendor access, and agent tool use.

DAR gives leadership, platform teams, security teams, and product teams a shared surface area. It turns "this is risky" into "this is the decision category, this is the boundary, this is the curated action, and this is how authority collapses."

Decision Category

A class of consequential decisions that share similar authority, risk, and governance requirements.

Examples include production deployment, break-glass access, data export, identity minting, vendor integration, and agent tool use. A decision category is useful because it lets the organization govern patterns of authority rather than isolated requests.

Decision categories matter because they help a CISO prioritize structural governance. The organization doesn't need to govern every action first. It needs to identify the few categories where authority, blast radius, irreversibility, and automation make the consequences highest.

Decision-Space Intelligence

A connected understanding of how authority, reachability, workflow, trust, and adoption actually operate inside an environment.

Decision-space intelligence is produced by connecting data from identity systems, vulnerability tools, cloud platforms, CI/CD systems, SaaS integrations, ticketing systems, incident records, access reviews, and business workflows. It shows where authority originates, where it expands, where paved roads are bypassed, and where unsafe trajectories remain reachable.

Disconnected dashboards rarely explain how work actually happens. Decision-space intelligence gives leaders evidence for where to build, constrain, deprecate, or invest next.

Mean Time to Authority

The time it takes for a team, workflow, system, or agent to obtain bounded authority through the paved road.

Mean Time to Authority measures whether the governed path is fast enough to beat unsafe workarounds. It should be reported the way operators experience it, often as median and P90, not only as an average.

Bounded authority must be usable under pressure. If the paved road is slower than a standing role, exception ticket, or legacy shortcut, teams will route around it and trust inflation will return.

Mean Time to Structural Closure

The time from discovering a reachable unsafe path to shipping the constraint patch that deletes or materially bounds that path.

This isn't the same as mean time to remediate. A ticket can close while the unsafe trajectory remains reachable. Structural closure means the system changed: a boundary tightened, a curated action was added, a foundry rule changed, an off-road path was deprecated, or collapse was bound to workflow state.

Mean Time to Structural Closure measures whether the organization is learning in the system, not just learning in the postmortem.

Reachability Ratio

A metric that estimates the maximum surface area an entity can affect if its authority is misused.

The entity may be a human identity, workload identity, service account, CI/CD runner, SaaS integration, vendor token, or agent. A simple version compares reachable assets to total assets within a defined scope. More mature versions weight reachability by sensitivity, environment, authority type, and consequence.

Reachability Ratio gives blast radius a measurable form. It helps show whether replacing standing authority with purpose-bound, collapsing authority is actually reducing what can be affected.

Security Leverage Ratio

A metric that measures how many governed paved-road executions occur for every manual security intervention required.

A simple version is:

governed paved-road executions ÷ manual security interventions for the same decision category

Security Leverage Ratio shows whether security is breaking the linear cost curve. In a posture-heavy model, more business activity creates more reviews, alerts, tickets, and analyst burden. In a structural model, more work should flow through governed pathways without proportional growth in manual security review.

Sovereign CISO

A CISO responsible for governing decision authority, not merely approving exceptions or sponsoring detection and response.

A sovereign CISO is the executive steward of the control-plane constitution. The role isn't to approve every consequential decision. The role is to ensure that the systems around those decisions enforce bounded authority, valid progression, and reliable collapse.

Trust Inflation Trend

A directional metric that shows whether broad, persistent, reusable trust is increasing or decreasing as the organization scales.

Trust Inflation Trend should be segmented by category: human authority, workload authority, agent authority, supplier or vendor authority, and cross-boundary authority. The goal of sovereign security isn't only to add new controls. It is to replace cheap trust with purpose-bound, collapsing authority and reduce the number of unsafe trajectories the organization can reach.

Category 6: Autonomy and Agent Terms

Agentic System

A system that can reason over context, choose among available actions, invoke tools, and progress toward a goal with some degree of autonomy.

Agentic systems matter because they turn authority into motion. A broad permission that once depended on careful human use can become an executable path when attached to an agent. The risk isn't only model output; it is what the agent can do with the authority and tools available to it.

Autonomous System

A system that can act without continuous human direction, often by interpreting inputs, selecting actions, invoking tools, and updating state.

An autonomous system doesn't need to be fully independent to create security risk. Even partial autonomy can matter if the system acts at machine speed, holds authority, touches sensitive data, or can create external consequences. The more consequential the action, the more governance must move before execution.

Autonomy

The ability of a system to act, decide, or progress through tasks without requiring a human to approve every step.

Autonomy removes the pause that once made weak governance survivable. Human operators often hesitate, apply context, or notice when a technically valid action feels wrong. Autonomous systems execute inside the decision space they are given.

Compositional Drift

The condition where risk appears in the sequence of actions rather than in any single action.

An agent may read data, draft output, invoke a tool, update state, and take external action, with each step authorized on its own. If those valid steps can combine into an unsafe outcome, the organization is dealing with a trajectory problem, not a simple access-control problem.

Delegated Identity

A pattern where an agent or automated system inherits a user's token, session, role, or permissions and can therefore act with the authority of that human or legacy identity.

Delegated identity is attractive because it is fast to implement. It is dangerous because it confuses who configured the agent with what the agent should be allowed to decide. A narrow task can inherit broad power.

Prompt Injection

A manipulation of an agent's instructions, retrieved context, input data, or tool environment that causes the agent to act contrary to the intended task or policy.

Prompt injection becomes more dangerous when the agent has broad authority. The problem isn't only that the model may produce a wrong answer. The problem is that a manipulated model may exercise real authority through tools, workflows, or control-plane access.

Speed Threshold

The point where execution moves faster than review, triage, or approval can meaningfully intervene.

When a system acts before oversight can matter, a human may still investigate later, but they are no longer on the critical path of the decision. The constraint has to exist at the point of execution.

Verified Context

Context that has been validated by a system of record, workflow state, policy check, broker, or approval path rather than merely inferred by the model.

Verified context is important because agents can misunderstand, hallucinate, or trust poisoned inputs. High-impact actions should depend on the context that the system can verify: the incident is active, the approval exists, the user is in scope, the artifact is signed, the dataset is authorized, or the workflow is in the correct state.

Workflow-Minted Authority

Authority is created for a specific workflow state rather than inherited from a standing role or broad identity.

Workflow-minted authority is a safer pattern for agents and automation. The system mints authority only when the workflow is in the right state, the purpose is clear, the scope is bounded, and the terminal condition is defined. When the workflow ends, the authority collapses.

Category 7: Implementation and Technology Anchor Terms

Admission Controller

A control-plane mechanism that intercepts requests before they become accepted system state.

In Kubernetes, admission controllers intercept requests to the API server after authentication and authorization but before the object is persisted. This makes them a useful example of inline governance: unsafe configurations can be rejected before they become a cluster state.

Cedar

An authorization policy language and engine used to express and evaluate access-control decisions.

Cedar is useful as an implementation anchor for policies that define whether an actor may perform an action on a resource. In trajectory governance, a Cedar-like policy layer is most powerful when combined with workflow state, purpose, and collapse, rather than used only as static permission checking. Cedar's official documentation describes it as a language for writing authorization policies and making authorization decisions based on those policies.

OIDC Federation

A federation pattern that allows systems to exchange identity claims and obtain credentials without relying on long-lived static secrets.

OIDC federation is often used to let CI/CD systems, workloads, or platforms obtain short-lived cloud credentials. In trajectory governance, it becomes valuable when those credentials are bound to workflow state, purpose, artifact identity, environment, and terminal conditions.

See also: Authority Foundry; Authority Minting; Dynamic Secret; Workflow-Minted Authority.

OpenFGA

An authorization system and modeling approach for fine-grained, relationship-based authorization.

OpenFGA answers authorization checks by evaluating relationships between users, objects, and an authorization model. It is useful as an implementation anchor for relationship-aware authorization, but relationship modeling alone doesn't govern unsafe trajectories unless it is combined with workflow state, purpose, sequence integrity, and collapse.

OPA / Rego

Open Policy Agent, or OPA, is a general-purpose policy engine. Rego is its policy language.

OPA is a useful implementation anchor for policy-as-code across applications, Kubernetes, CI/CD, API gateways, and other systems. Its own documentation describes it as a general-purpose policy engine that lets teams specify policy as code, while Rego is used to reason over structured data such as API requests, infrastructure-as-code, and configuration data.

Policy-as-Code

The practice of expressing policy decisions in code or machine-readable rules that can be tested, versioned, reviewed, and enforced by systems.

Policy-as-code is useful for trajectory governance when it evaluates more than isolated permissions. It should help enforce valid states, required prerequisites, scoped authority, and progression rules. Tools such as OPA/Rego and Cedar are common implementation anchors for this pattern.

Provenance

Evidence about where an artifact came from, how it was built, and which process produced it.

Provenance is important for deployment governance because production authority shouldn't be granted merely because a pipeline has an artifact. The system should be able to verify that the artifact came from the expected source, build process, identity, and workflow. SLSA frames provenance and integrity as core parts of securing software artifacts and supply chains.

Relationship-Based Authorization

An authorization model where access is determined by relationships among users, objects, groups, organizations, resources, or other entities.

Relationship-based authorization can express richer access models than simple role checks. OpenFGA describes Relationship-Based Access Control as allowing access rules to depend on relations between a user, an object, and related objects. In this book's model, relationship authorization is useful but incomplete unless it also considers workflow state, purpose, sequence, and terminal conditions.

Secrets Broker

A system that issues, mediates, rotates, or revokes secrets and credentials on behalf of workloads, users, workflows, or agents.

A secrets broker can support authority minting when it issues credentials only under valid workflow conditions. It becomes more powerful when combined with purpose, scope, time limits, and automatic collapse rather than simply distributing secrets on demand.

SLSA

Supply-chain Levels for Software Artifacts, a framework for improving the integrity of software artifacts and the systems that build them.

SLSA is useful in this book as an implementation anchor for artifact integrity, provenance, and build-system trust. A deployment paved road may require SLSA-aligned provenance before production deployment authority can be minted. SLSA describes itself as a security framework of standards and controls to prevent tampering, improve integrity, and secure packages and infrastructure.

SPIFFE / SPIRE

SPIFFE is a framework and set of specifications for workload identity. SPIRE is an implementation used to issue and manage those identities.

SPIFFE/SPIRE is an implementation anchor for strongly attested workload identity. SPIFFE's documentation describes short-lived cryptographic identity documents called SVIDs that workloads can use for authentication, including TLS or JWT-based flows.

References and Further Reading

The works below informed the language, context, and technical foundations used throughout this book. This is not intended to be an exhaustive bibliography of cybersecurity, software engineering, or AI governance. It is a selected set of sources that support the book's central themes: security engineering, structural constraint, authority governance, control-plane security, software supply chain integrity, agent authorization, organizational adoption, and the transition from posture to trajectory governance.

Security engineering and secure-by-design foundations

Anderson, Ross J. *Security Engineering: A Guide to Building Dependable Distributed Systems*. 3rd ed. Indianapolis: Wiley, 2020.

Beyer, Betsy, Ana Oprea, Heather Adkins, and Piotr Lewandowski, eds. *Building Secure and Reliable Systems: Best Practices for Designing, Implementing, and Maintaining Systems*. Sebastopol, CA: O'Reilly Media, 2020.

McGraw, Gary. *Software Security: Building Security In*. Boston: Addison-Wesley, 2006.

Cybersecurity and Infrastructure Security Agency. *Shifting the Balance of Cybersecurity Risk: Principles and Approaches for Secure by Design Software*. Washington, DC: CISA, 2023.

National Institute of Standards and Technology. *Secure Software Development Framework (SSDF) Version 1.1: Recommendations for Mitigating the Risk of Software Vulnerabilities*. NIST Special Publication 800-218. Gaithersburg, MD: NIST, 2022.

Saltzer, Jerome H., and Michael D. Schroeder. "The Protection of Information in Computer Systems." *Proceedings of the IEEE* 63, no. 9 (1975): 1278–1308.

These sources are strong anchors for the book's design philosophy because they connect security to systems design rather than after-the-fact inspection. CISA's Secure by Design guidance is especially useful for the book's claim that responsibility must move into product and platform design, while NIST SSDF supports the secure software development thread.

Risk governance, posture, vulnerability management, and zero trust

Cybersecurity and Infrastructure Security Agency. *Known Exploited Vulnerabilities Catalog*. Washington, DC: CISA.

FIRST. *Common Vulnerability Scoring System Version 4.0: Specification Document*. Forum of Incident Response and Security Teams, 2023.

MITRE. *MITRE ATT&CK Enterprise Matrix*. McLean, VA: MITRE.

National Institute of Standards and Technology. *Cybersecurity Framework 2.0*. NIST Cybersecurity White Paper 29. Gaithersburg, MD: NIST, 2024.

National Institute of Standards and Technology. *Risk Management Framework for Information Systems and Organizations: A System Life Cycle Approach for Security and Privacy*. NIST Special Publication 800-37, Revision 2. Gaithersburg, MD: NIST, 2018.

Rose, Scott, Oliver Borchert, Stu Mitchell, and Sean Connelly. *Zero Trust Architecture*. NIST Special Publication 800-207. Gaithersburg, MD: NIST, 2020.

Chandramouli, Ramaswamy. *A Zero Trust Architecture Model for Access Control in Cloud-Native Applications in Multi-Cloud Environments*. NIST Special Publication 800-207A. Gaithersburg, MD: NIST, 2023.

These references support the book's critique of posture and deterministic severity without dismissing current security practice. CVSS, KEV, ATT&CK, NIST CSF, RMF, and zero trust are the mainstream context against which the book's trajectory-governance argument becomes clearer.

Control planes, policy-as-code, and structural enforcement

Amazon Web Services. *AWS Organizations: Service Control Policies.* AWS Documentation.

Amazon Web Services. *Amazon Verified Permissions Documentation.* AWS Documentation.

Cedar Policy Language. *Cedar Policy Language Reference Guide.*

GitHub. *About Protected Branches.* GitHub Documentation.

HashiCorp. *Vault Documentation.* HashiCorp Developer.

Kubernetes. *Admission Control in Kubernetes.* Kubernetes Documentation.

Open Policy Agent. *Policy Language.* Open Policy Agent Documentation.

These are useful implementation anchors for Chapters 8 and 9. Kubernetes admission controllers, AWS SCPs, OPA/Rego, Cedar, GitHub branch protection, and Vault-style secrets brokering all illustrate the same pattern: moving security closer to the place where intent becomes state, authority, or executable action.

Workload identity, authorization, and relationship-based access

OpenFGA. *OpenFGA Documentation.* OpenFGA.

Pang, Ruoming, Ramon Caceres, Mike Burrows, Zhifeng Chen, Pratik Dave, Nathan Germer, Alexander Golynski, Kevin Graney, Nina Kang, Lea Kissner, Jeffrey L. Korn, Abhishek Parmar, Christina D. Richards, and Mengzhi Wang. "Zanzibar: Google's Consistent, Global Authorization System." In *2019 USENIX Annual Technical Conference.* Renton, WA: USENIX Association, 2019.

SPIFFE. *SPIFFE Concepts.* SPIFFE Documentation.

SPIFFE. *SPIRE Concepts.* SPIFFE Documentation.

These references support the authorization discussion in Appendix D. Zanzibar and OpenFGA are important because they show how relationship-based authorization can scale, while SPIFFE/SPIRE provide the workload-identity foundation for purpose-bound and brokered authority.

Software supply chain, provenance, and trusted delivery paths

Open Source Security Foundation. *Supply-chain Levels for Software Artifacts.* SLSA.

OpenSSF. *SLSA Specification.* Open Source Security Foundation.

These references support the book's production-deployment and "governing the governor" sections. SLSA is especially relevant to artifact provenance, signed build paths, and the idea that the governance plane itself must be treated as a high-integrity supply chain.

Paved roads, golden paths, and platform engineering

Backstage. *Backstage Software Catalog and Developer Platform*. Backstage.

Google Cloud. "Golden Paths for Engineering Execution Consistency." Google Cloud Blog, 2023.

Netflix Technology Blog. "The Show Must Go On: Securing Netflix Studios at Scale." Netflix Technology Blog, 2021.

Spotify Engineering. "How We Use Golden Paths to Solve Fragmentation in Our Software Ecosystem." Spotify Engineering Blog, 2020.

These references support the paved-road argument in Chapters 6, 8, 10, and 11. The Spotify and Netflix sources are particularly useful because your book relies on the idea that the supported path must be usable, trusted, and adopted under pressure, not merely documented.

AI, agents, and emerging authorization problems

Anthropic. "Introducing the Model Context Protocol." Anthropic, 2024.

Cloud Security Alliance. *AI Controls Matrix*. Cloud Security Alliance, 2025.

Model Context Protocol. *Authorization*. MCP Documentation.

Model Context Protocol. *Security Best Practices*. MCP Documentation.

National Institute of Standards and Technology. *Artificial Intelligence Risk Management Framework (AI RMF 1.0)*. NIST AI 100-1. Gaithersburg, MD: NIST, 2023.

National Security Agency. *Model Context Protocol: Security Design Considerations for AI-Driven Automation*. Fort Meade, MD: NSA, 2026.

OWASP. *OWASP Top 10 for Large Language Model Applications*. OWASP GenAI Security Project, 2025.

These references are important for the agent and autonomy material. The MCP, OWASP, NIST AI RMF, CSA AI Controls Matrix, and NSA MCP guidance give readers a current source trail for prompt injection, tool use, authorization, excessive agency, context handling, and agent-driven automation risk.

Systems thinking, safety, and constraint design

Leveson, Nancy G. *Engineering a Safer World: Systems Thinking Applied to Safety*. Cambridge, MA: MIT Press, 2012.

Leveson, Nancy G., and John P. Thomas. *STPA Handbook*. Cambridge, MA: MIT, 2018.

Meadows, Donella H. "Leverage Points: Places to Intervene in a System." The Sustainability Institute, 1999.

These sources help situate trajectory governance in a broader systems-thinking tradition. Leveson's work is especially relevant to the idea that safety in complex systems is a control problem, not merely a component-failure problem. Meadows is useful for the book's recurring point that the highest-leverage interventions are often changes to system structure, rules, and information flows.

Organizational change, adoption, and choice architecture

Christensen, Clayton M., Taddy Hall, Karen Dillon, and David S. Duncan. *Competing Against Luck: The Story of Innovation and Customer Choice*. New York: HarperBusiness, 2016.

Lafley, A. G., and Roger L. Martin. *Playing to Win: How Strategy Really Works*. Boston: Harvard Business Review Press, 2013.

Moore, Geoffrey A. *Crossing the Chasm: Marketing and Selling Disruptive Products to Mainstream Customers*. 3rd ed. New York: HarperBusiness, 2014.

Rogers, Everett M. *Diffusion of Innovations*. 5th ed. New York: Free Press, 2003.

Thaler, Richard H., and Cass R. Sunstein. *Nudge: The Final Edition*. New York: Penguin Books, 2021.

These are the strongest references for Chapters 6 and 11 and Appendix E. Thaler and Sunstein support the choice-architecture framing; Rogers and Moore support the adoption and institutionalization argument; Lafley and Martin support the "where to play / how to win" transition logic. Christensen and colleagues support the Jobs-to-be-Done framing used in Appendix I to argue that security tooling should design around what developers, operators, and agents are trying to accomplish.

Financial and market-systems analogies

Securities and Exchange Commission. "Risk Management Controls for Brokers or Dealers with Market Access." Rule 15c3-5. Washington, DC: SEC, 2010.

This reference supports the Chapter 6 analogy to pre-trade risk controls. The financial-systems analogy is narrow but useful: in automated, high-speed environments, certain controls must operate before market access creates harm, rather than relying only on post-event detection and correction. SEC Rule 15c3-5 is the specific regulation that requires broker-dealers to implement such pre-trade controls, and is the source most closely aligned with the analogy as the chapter uses it.

Trust, platforms, and trust inflation

König, René, and Patrick Sumpf. "User Loyalty Unbroken? Inflation of Trust in Digital Platforms." *Global Perspectives* 2, no. 1 (2021).

This reference is useful because the manuscript already distinguishes your use of "trust inflation" from König and Sumpf's use of "inflation of trust" in digital platforms. Keeping this citation makes the distinction credible and avoids making the term appear ungrounded.

Index

Page numbers reference the printed page numbers shown in the running headers.

A

B

C

D

R

S

About the Author

Himanshu Tiwari is a product leader focused on cybersecurity with nearly two decades of experience across product management, consulting, and large-scale organizational transformation as an operator, advisor, and builder. His work has focused on building industry-shaping products and helping enterprises design security programs and platforms.

He has held leadership roles across cybersecurity and technology organizations, including Cigital, Synopsys, Diebold Nixdorf, and Amazon. He is also a recognized voice and public speaker on product leadership and cybersecurity, with a focus on how security organizations can move beyond control enforcement toward systems, incentives, and products that help businesses move faster while managing risk.

His perspective is shaped by systems thinking: the belief that security outcomes are not created by controls alone but by how authority, incentives, nudges, workflows, data, and engineering decisions interact over time. Beyond Posture reflects his conviction that modern security must move beyond static views of risk and begin governing the trajectories systems are allowed to take.

www.ingramcontent.com/pod-product-compliance
Lightning Source LLC
LaVergne TN
LVHW070826220926
848486LV00027B/18